On-Chip Instrumentation

Neal Stollon

On-Chip Instrumentation

Design and Debug for Systems on Chip

 Springer

Neal Stollon
HDL Dynamics, Dallas
TX, USA
neals@hdldynamics.com

ISBN 978-1-4899-9230-7   ISBN 978-1-4419-7563-8 (eBook)
DOI 10.1007/978-1-4419-7563-8
Springer New York Dordrecht Heidelberg London

Printed on acid-free paper

Springer is part of Springer Science+Business Media (www.springer.com)

Preface

When I started this book, I thought I understood the world of on-chip debug–after all, I had been part of one of the leading startups in the area for 5 years and had been a participant in a number of standard and industry organizations that were leading the world of on-chip debug and instrumentation into the next wave. As I gathered my materials, I grew more impressed by the day and by the month at the body of work that this topic has accumulated, in industry and in academia, in every nook and cranny of the embedded systems business, from embedded processor, to bus architecture, to FPGA, to IP development; engineers have developed and customized a truly impressive range of on-chip debug and instrumentation solutions to address and support their products and to enable an increasingly capable infrastructure that does much more than the prosaic word debug implies and starts to address the full potential of what on-chip instrumentation can truly provide for the electronics industry.

This book came about, in part, because of the lack of a comprehensive discussion of on-chip debug instrumentation. This seems to have been an area where the experts come about from on-the-job experience and in ad hoc methods. On-chip debug is an integral part of most modern processor and system on-chip (SoC) design, but in my experience it is not a topic given in-depth discussion in engineering school (universities take note). Most engineers' experience of on-chip debug is limited to plugging into the JTAG port and running the software, with little understanding of what goes on within. This text tries to provide a general overview of the different types of on-chip debug that goes into a design.

This book is structured into three main sections; the first, Chaps. 1–7, is an introduction to the variety of concepts that make up on-chip debug, in particular looking at some of the history and well-established infrastructure, including an overview of JTAG from a debug, rather than test, point of view. It also looks at aspects of processor- and bus-level instrumentation and discusses multicore on-chip debug issues The second section, Chaps. 8–11, addresses a number of the standards and industry efforts that are ongoing in areas ranging from instrument interfaces to JTAG advances, some of which, like Nexus and OCP-IP, I have been involved in, and others that have been a learning experience for me over the last year, all of which I believe will form the core basis for the next generation of on-chip debug. The third section, Chaps. 12–15, is a survey of some of the wide variety of commercially supported solutions for on-chip debug, addressing a limited cross section of the types of on-chip instruments that are available for different processors and SoCs.

Some areas related to on-chip debug have been intentionally kept generic and out of the discussion to maintain the focus on the on-chip instrumentation. Notably, I have kept any detailed discussion of probes and host-based debugger software to a minimum, other than what is required to make the concepts of JTAG and trace understandable. This may seem unusual, but the reasons for this are two-fold. First, the topic of debug probe and software design is at least a book in itself. Second, the commercial business involved in probes and debug software is a significant business unit for most processor companies as well as the dozens of companies that provide probe and software solutions (many run by people I know) that address the range of debug options. To mention any one example in any detail would ignore the rest that are equally deserving of mention.

Few are of variety of instrumentation- and debug-related areas I cover are discussed exhaustively. This is due to both limitations on space and a large amount of supplemental detailed information available elsewhere for those who want to explore in more depth. Similarly, I have intentionally avoided discussion of some of the more advanced implementations, in order to keep the text accessible to a more general reader. For virtually all topics, I highly recommend the reader to directly contact the IP or chip vender or standards group for more detailed and updated information on the topics. Those interested in instrumentation products can find an amount of online resources that address specific instrumentation solutions in minute detail. The amount of documentation avaliable on MIPS EJTAG or ARM ETM, for example, can put page length of *War and Peace* to shame.

The standards-related activities are somewhat less well documented, in some cases because they are work in progress. However, there is a lot of follow-on information out there for those who search. So I have tried to focus on what I think are the interesting or unique parts of different instrumentation solutions, with the assumption that readers interested in more detail can find it.

I want to acknowledge a number of people in the industry who have helped me along the way, especially Rick Leatherman and the on-chip instrumentation team of the First Silicon Solutions group at MIPS, who got me started in thinking about on-chip instrumentation and who taught me far more they realize about on-chip debug technologies and the businesses involved. I also thank the current and past members of the Nexus IEEE 5000 Forum and members of the OCP-IP Debug Working Group, with special recognition to Bob Uvacek, my longtime compatriot in the working group.

Last, but by far not least, I want to acknowledge my family, without whom I am nothing. My wife Marcy, my daughters Courtney and Naomi, my son Eric, and my mom Rita Bickel Stollon (of blessed memory) were patient and understanding of the time I spent working on this book. Finally, I dedicate this book to my family but especially to my father Arthur Stollon (of blessed memory), who proofread everything I wrote while I was in school and taught me "be prepared to trudge thru the wilderness to get a change at the limelight".

Dallas, TX   Neal Stollon

Contents

Chapter 1
Introduction

With each new generation of digital system-on-chip (SoC) technology, the level
of integration, functionality, and complexity provided on a single chip has increased
significantly. A problem that goes hand in hand with this increased amount of integration and functionality is that analysis tasks and difficulty associated with getting
a design working and integrated increase at least proportionally to the size and
complexity of the chip. Over a range of SoC types – ASIC, ASSP, system FPGAs,
or a dozen other variants and platforms – there is a common need for better debug
solutions. As more processing elements, features, and functions are simultaneously
being embedded in the silicon, the emerging level of embedded complexity outstrips
the capability of stand-alone logic analyzer-, debugger-, and emulator-based diagnostic tools for embedded designs. Although these tools allow the capture of data
off the system data bus, they work only as long as every access (read and/or write)
occurs over the external data bus. This issue points to a growing gap in terms of
effectively being able to provide the necessary controllability and, in particular, the
visibility of the internal operations of a complex system.

1.1 The Need for On-Chip Debug

The need for improved methods of observing and analyzing embedded processor and
SoC operation has increased at a pace at least proportional to the explosive growth
in electronic system designs and new intellectual property (IP) cores that populate
them. The analysis side of the SoC world is then forced into a constant process of
catching up to the designer's ability to add cores and integrate new resources on
chips. With an ever-shortening development cycle, and often several generations of
products produced in parallel or rapid succession, standardized embedded tools and
capabilities that enable quick analysis and debug of the embedded IP are a critical
factor in keeping SoC verification a manageable part of the process.

Most engineers involved in complex design will agree that verification and validation have become a critical stumbling blocks in the development and release of new
devices. This is now equally true of the software components of those systems as well
as the hardware. Better ways to address the verification and analysis of complex SoC

N. Stollon, *On-Chip Instrumentation: Design and Debug for Systems on Chip*,
DOI 10.1007/978-1-4419-7563-8_1, © Springer Science+Business Media, LLC 2011

designs, and corresponding new methodologies, tools, and capabilities, are needed to get the job done at different phases of the design and development cycle.

For brevity and convenience, we adopt some general definitions in this book. Analysis relating to pre-silicon design stages, in particular Electronic Design Automation (EDA) tools and methods and their use in RTL (register transfer level) and ESL (electronic system level), are collectively referred to as *verification*. Analysis relating to in-silicon analysis (sometimes referred to as post-silicon analysis), and in particular, tasks related to the use of on-chip instrumentation, are collectively referred to a *debug* (Fig. 1.1).

In-silicon debug provides complementary alternative methods to digital simulation as a means of viewing and analyzing embedded signals. Simulation, although a critical factor in verification, is not a total verification solution for embedded SoC. Simulation alone cannot address all the facets and nuances of physical hardware. In addition, it is not realistic to simulate large multiprocessor architectures for the extremely large numbers of cycles required to evaluate the software-specific aspects of system operation and real-world system performance. Although accelerated simulation, co-simulation, and emulation environments provide a stop-gap method of improving the simulation capability in observing system performance, these often introduce costs and complexities beyond the resources of many projects. On-chip instrumentation and debug approaches have evolved as a low-cost and efficient alternative of increasing system visibility which focus on the actual final hardware product rather than its model.

System-on-chip debug, like most verification philosophies, seeks to maximize test functionality and ease of verification while reducing the overall end-user cost. There is a constant trade-off that must be made on resources dedicated to system analysis and debug versus the system cost of including these features. The value of debug is mainly perceived during the development cycle (hardware or software), where operational questions and integration issues for the key processing blocks are unresolved. After the system is "fully debugged," the hardware investment in debug capabilities becomes much more application-focused. In the past, this often led to the removal of debug features (and associated gates and pins) in order to gain a small reduction in die size. These trade-offs have shifted in recent years by dramatic

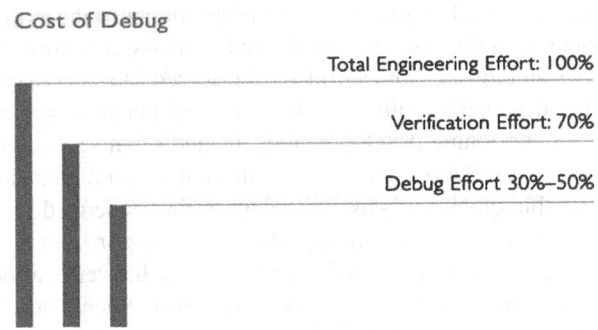

Fig. 1.1 The costs of debug

increases in complexity; gate availability and overall speed of system operation have changed the norm in chip design from a "core and gate-limited" to a "pin and IO-limited" focus. The debug question has migrated from how many gates can be spent on debug to an ease-of-use and bandwidth issue: How much of the system resources and IO is needed to debug the system successfully?

1.2 Instrument- (**in-silicon) and EDA- (Presilicon) Based Verification

EDA (electronic design automation) flows that include the tools and methods for modeling, simulation, and analysis of SoC receive much attention and have evolved a variety of solutions to address verification needs for pre-silicon design, from diverse simulation-based methodologies to emerging formal and assertion-based methods and, increasingly, system-level abstraction. This verification flow largely works under the assumption that the verification effort is essentially completed when the design files are handed off to the foundry for fabrication. Anyone who has been involved in the in-silicon debug cycle, loosely defined as everything that must be verified and integrated from the time silicon is received back from the foundry to the point of being ready for a production release, knows that this is far from the case.

Although improved tools and rigor in pre-silicon verification are essential and play a important role in getting to working first-pass silicon, the use of in-silicon debug has received much less attention. However, as we discuss, in-silicon debug plays an essential role in addressing full-speed testing in real environments. It allows for more exacting analysis of interactions too subtle for models in simulation to address, such as unforeseen environmental variables, external constraints, etc. Analysis, and bug-fixing, including resolving hardware/software issues that cannot easily be addressed at speed other than by analysis of the in-silicon hardware platform itself.

A design team, in addition to having to address the issues of verification and model reuse at different stages of the design flow, typically must also develop debug flows to address both hardware prototypes and in-silicon verification of both the hardware and software in the system when getting chips to market.

With larger and more complex chip architectures and designs, supporting larger and more complex software applications, the penalties of discontinuities between pre-silicon verification and in-silicon debug are increasing. More consistent and common environments reduces costs and trade-offs of getting silicon not just designed but working and out the door to the end customers (Fig. 1.2).

The cost of debug in the development of complex SoC systems has not received the level of analysis that other parts of the methodology, such as EDA tools, have, but it is pragmatically known to be a significant portion of the overall cost of releasing new systems. EDA tools and flows have focused on evolving a variety of solutions to address pre-silicon verification, with diverse simulation-based methodologies that leverage high-level verification languages and formal and assertion-based

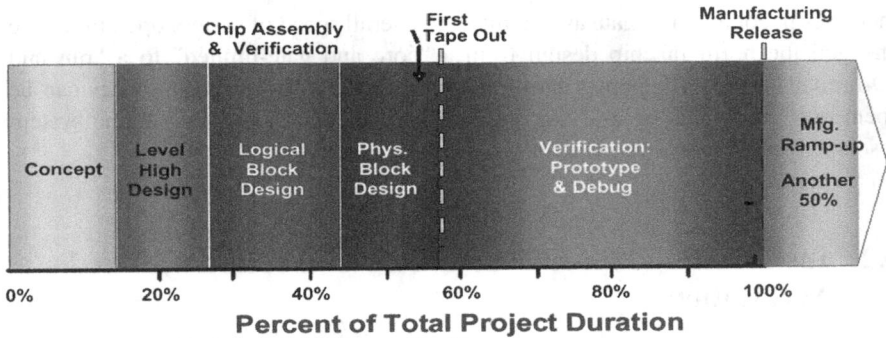

Fig. 1.2 Debug duration over a project

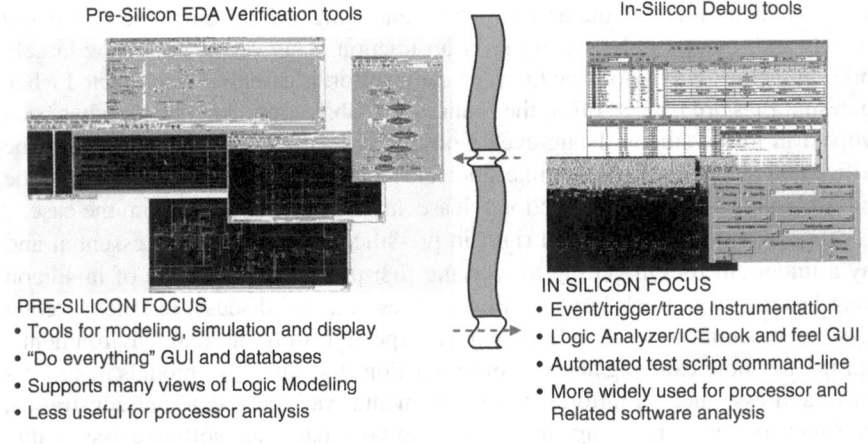

Fig. 1.3 Pre-silicon vs. in-silicon analysis

methods to verify at increasing levels of system-level abstraction. As we discuss, these are complementary to instrumentation-based debug (Fig. 1.3).

Debug typically relies on a toolbox of methods, instrumentation IP, and tools to support analysis of hardware-based systems and their software applications. Debug includes combinations of software tool methods (which can be thought of as "print" statements and breakpoints) and hardware methods (monitoring of events using Instrumentation IP to capture information for display and analysis).

Successful in-silicon analysis of next-generation systems will rely as much on system-level thinking in leveraging and reuse of verification efforts done during the pre-silicon verification cycle as on specific debug instrumentation approaches in providing closure to address the SoC verification and analysis problem (Fig. 1.4).

Complex architectures have spurred the requirement for new methodologies and capabilities to address the analysis and instrumentation needs of these architectures. We are arguably moving toward a new inflection point of requiring a sea change in debug assumptions, based on changing design methodologies that widely embrace

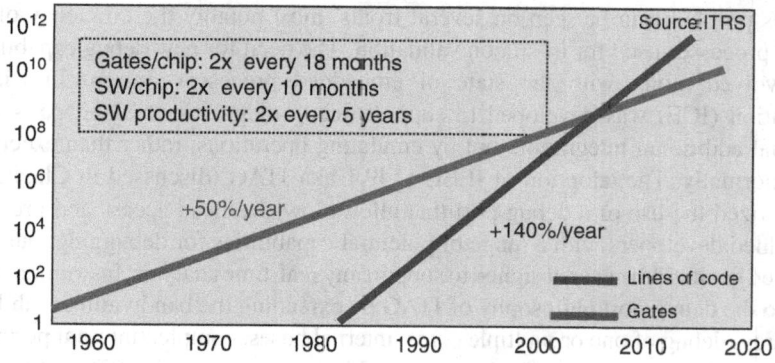

Fig. 1.4 Growth of hardware vs. software complexity

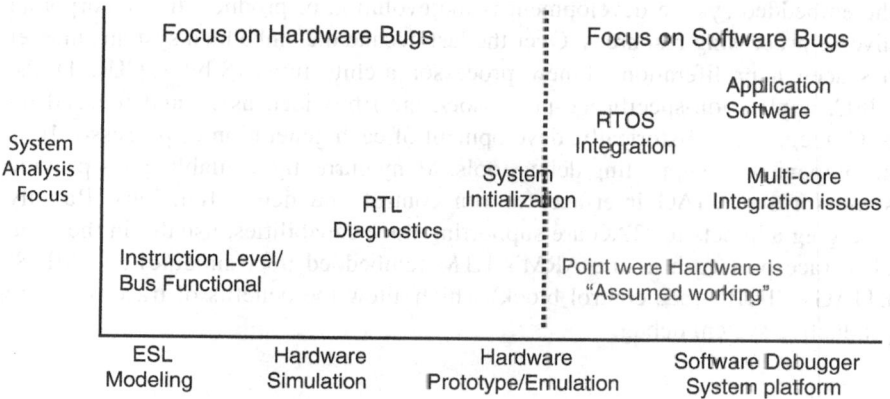

Fig. 1.5 Verification abstraction vs. debug tasks

multiprocessor architectures and their associated software development and integration issues, dramatically increased gate count availability, and increased complexity in all the diverse interfaces and peripherals making up a SoC device.

Looking historically at the major inflection points for EDA verification, debug tools, and silicon complexity in Fig. 1.5, it is interesting to note the interwoven relationship between these different but closely related technologies that are central to the progress of many aspects of the evolution of leading-edge electronics technology. The emergence of new EDA tools is both a driver and a result of new and increasingly complex levels of systems architectures. Similarly, complexities in architectures have spurred the requirement for new debug methodologies and capabilities to address the needs of these architectures. We are arguably in the middle or moving toward a new inflection point, based on changing design methodologies that widely embrace multiprocessor architectures, dramatically increased gate count availability, and increased complexity in all the diverse interfaces and peripherals making up an emerging SoC device.

This problem can be seen on several fronts, most notably the efficiency of the debug processes used for in-silicon validation. The need for new debug capabilities has evolved along with the state of embedded processor design. In Circuit Emulation (ICE) was developed to support debug of processor-based parts with minimal additional integration, but by emulating operations, rather than executing them normally. The adoption of IEEE 1149.1 aka JTAG (discussed in Chapter 3) popularized the use of a debug port that allows low overhead access and provides embedded developers with a range of potential capabilities for debugging, but with a limited bandwidth never designed to support any real-time analysis. Instrumentation adds to the debug port philosophy of JTAG by extending the bandwidth capabilities to address debug of one or multiple cores, internal buses, complex internal peripherals, and high-speed data traffic found at SoC levels of complexity (Fig. 1.6).

Looking at on-chip instrumentation in the proper context, it is useful to examine more traditional embedded systems debug. The ever-increasing trend in the embedded system development is the evolution of products that incorporate diverse processing resources. Over the last decade the embedded systems market has seen a proliferation of new processor architectures (8-bit MCUs, DSPs, RISC, application-specific co-processors, etc.) provided as IP and focused on SoC integration. Historically, development of each generation of processor IP is accompanied by supporting debug tools. Many currently available cores provide some form of JTAG interface for run control and debug functions. Rapidly emerging adjuncts to JTAG are supporting trace capabilities, usually in the form of a trace port such as the ARM's ETM (embedded trace module) and MIPS' EJTAG + TCB (trace control block) which allow the benefits of trace for more productive system debug.

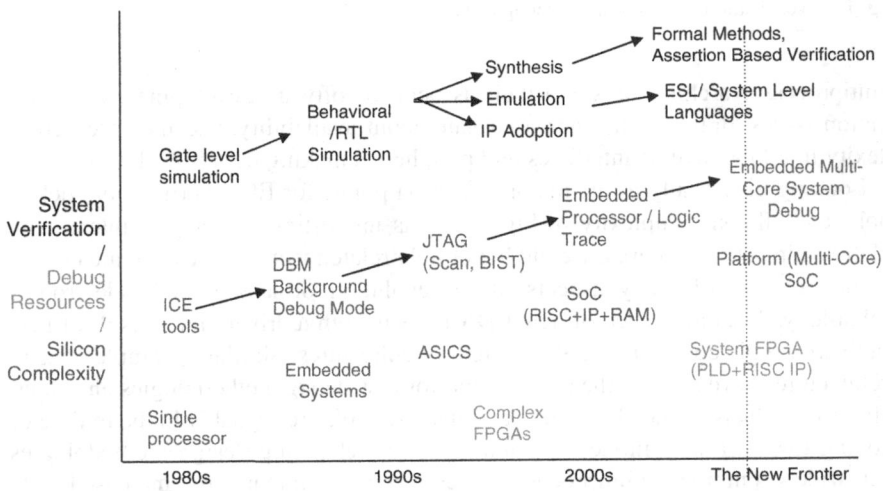

Fig. 1.6 SoC verification, debug, and complexity

1.3 SoC Debug Requirements

Analysis, at all levels of implementation, relies on methods of configuration, control, and data capture. Control refers to the manipulation of a system, outside of its normal execution, for the purpose of debug, analysis, and verification. Control can be influenced at any time during execution of the system, so it is a "real time" method in this respect. A simple control example might be to execute a single instruction, but more typically it may involve execution of a range or duration of operations tied to an initializing event and a concluding event. Configuration is actually nothing more than a special case of control, referring to the initial setup of a system to a known state. In some cases, this configuration may be part of the normal execution of a system (for instance, the default settings used after reset). Examples include setting of mode or configuration bits for cores, arbitration states for buses, and loading of data into specific locations in the system to (re)produce a system state for a particular operation or sequence of interest. Data capture refers to the export and storing of some system information occurring at a user-defined time. A simple example would be capture of a register value occurring at some triggering event in the system. Both ESL and instrumentation tools have similar requirements at a SoC level in how to address these tasks, and ideally would rely on standardized mechanisms for implementing them.

Most SoC include a programmable processor and in many cases, multiple processors as the core functionality. They also consist of infrastructure, either in terms of dedicated coprocessors or other logic and a communications infrastructure to allow both intercore and chip-to-outside world communications. The analysis of processors and the rest of a complex chip follow different paths and have traditionally relied on different approaches to verification and debug. Digital hardware design, on the contrary, typically relies on a synthesizable RTL model that assumes implicit clock-cycle timing during simulation. RTL has been the primary debug tool for configuration, control, and data capture of dedicated logic-based portions of the architecture, with hardware support based on either on- or off-chip logic analysis, although with the advent of synthesizable ESL language subsets and methodologies, these functions may be absorbed into the ESL of design flow. In either case, merging simulation and synthesis approaches have been proven in countless designs over the last 15 years, because logic-based functions can typically be analyzed over the range of fewer than a million clocks cycles, which is manageable for both simulation and logic analysis. Conversely, processor architectures, while relying on synthesis for implementation, are less successful in using RTL and logic analysis approaches due to the length of time required for execution of complex algorithms, and complexities of hardware and software interactions that are not amenable to RTL simulation and related approaches.

Simulation is always an important part of the development flow; just as important is the ability to analyze hardware during prototyping and system verification and, increasingly, on the final products themselves. Although the focus of much of the verification world has been on simulation-based verification technologies, instrumentation provides a counterpoint that focuses on the physical hardware. The problem in analyzing embedded information (on-chip processors and buses) in hardware in many cases devolves to a visibility problem – *it is difficult to fix problems that you cannot see.* The test community traditionally has referred to this problem in terms of levels of controllability and observability of a design. It is important to note that this analysis visibility cannot be adequately addressed by traditional on-chip test methods such as JTAG scan, and that the analysis and instrumentation problem, while overlapping with test issues and techniques in many cases, is fundamentally different.

One method of working around these analysis bottlenecks in simulation is to build hardware emulators or prototypes of (usually field-programmable) hardware implementations of the digital and possibly analog portions of a system. These hardware systems will run orders of magnitude faster than simulations, making running software applications feasible, but they are still typically at least an order of magnitude slower than the final silicon system, which results in both false positive (errors in the emulator that are not in silicon, due to differences in timing paths, synchronization of subsystems, etc.) and negative problems (found in the silicon that are not seen in the hardware, due to lower speed) and in many cases still not being able to run the system application at a speed compatible with the final system requirements.

Modern silicon system-level implementation typically proceeds through a design life cycle of increasing detail and refinement that must include modeling, verification, and analysis of hardware and software components. Software development has typically relied on analysis with a hardware target using ISS (instrument set simulation) models where timing is abstracted or nonexistent. These ISS models can vary significantly from vendor to vendor, which inhibits general methods for model compatibility between different core models and their integration with RTL simulation. RTL is synthesized into gate-level implementations that map into hardware and become a deliverable product, along with software that is either embedded as part of the product or added at a later stage by customers. More complex modeling is complicated in modern devices by several factors.

- Preferred software development environments may vary significantly for different processors. Although hardware development tools have developed in parallel but largely independent of different ways to implement a design (programmable logic, ASIC, ASSPs, and their related digital IPs), a limited number of common representations (gate-level, RTL, hardware itself) allow for straightforward integration; software development tools and models are developed by and in conjunction with processor and software IP vendors and have more limited commonality, for modeling and verification of multiple processors or even different processor targets running a common algorithm. The problem is even more acute for debug-related tasks, where different debuggers have different features. More commonality is found in use of GNU debuggers (GDB), versions of which have been developed by and for many processor architectures. GDB and variants are commonly used as a user interface for configuration, control, and data capture of software architectures during ISS, emulation, and in-silicon debug.

- For multicore devices, different suppliers often provide subsystems, in terms of hardware blocks, each developed with its own assumptions and incompatibilities in ISS modeling. Due to a lack of standardized sequential timing models in software languages used to develop ISS models (C, C++), new modeling approaches that include understanding of sequential and concurrent operations are needed to model systems that include multiple processors (running under different clocking and instruction flows) and processors and their supporting blocks (buses, peripherals, interfaces, etc.) that are typically modeled at RTL or other hardware methods.
- These may be in simulated SystemC, an ESL developed to support concurrent modeling of systems having processor architectures and software and supporting hardware blocks. SystemC combines compatibility with C++ as a class library with a set of corresponding modeling and simulation features similar to those used in RTL.
- Real device speeds are higher (typically by orders of magnitude) than that achievable by simulation. As a result, system modeling relies on abstractions and simplifications to increase simulation performance to a point where it is feasible to run software applications over the multimillion cycles needed to verify operation. Complexity and performance are further impacted if different subsystems are asynchronous or have other analysis-intensive incompatibilities. The lowest risk and often the simplest solution to real-time analysis is to use the actual hardware; however, even with added instrumentation, there are significant limitations in observability and controllability of a design as discussed earlier, so while hardware is a good verification platform, it is limited as an analysis platform. Simulation does not have the same limitations, because all signals are visible. One of the more important simulation efforts of SystemC is related to trade-offs between speed and visibility with TLMs (transaction-level models) that, by abstracting away noncritical functionality or timing, can simulate orders of magnitude faster than cycle-timed RTL models while being integrated with RTL models. Integration between TLM and RTL blocks in a simulation, while providing more resolution of signal analysis (at the expense of increased simulation timing), is still an area of active development.
- Complicating simulation analysis further is the modeling of the complex environments in which the device must operate. These can include the need for modeling a complex stimulus with both signal and noise characteristics, human interfaces, and analog subsystems that have their own modeling and analysis complexities, which are incompatible with large-system digital analysis and have their own traditional (frequency-domain-based) analysis methods. The effective integration of mixed analog and digital systems remains an open area of refinement in EDA analysis methods and in hardware-based debug and analysis; test features within ESL tools include the ability to model many analog and system characteristics as part of verification blocks (test benches) as well as the ability to integrate models from verification-level languages (Specman, Vera, Testbuilder, etc.) that have been developed and are being incorporated into new versions of RTL languages such as SystemVerilog.

Choosing among many design trade-off efforts is a tiered approach of modeling refinement and migration from ESL, to more detailed models, to hard platforms, to final silicon. As the modeling and analysis move from simulation to hardware, another factor to consider is an accompanying loss of visibility and access in the internal signal operation. In simulation, all signals, variables, and modeling parameters are available for viewing, and in most cases, for direct modification, providing a rich analysis environment, regardless of other limitations. Hardware, whether in emulation system or in final silicon, has limitations for debug purposes on the amount of visibility and control of embedded signals available at the system IO pins (Fig. 1.7). In this hardware environment, instrumentation significantly increases the amount of real-time visibility and control of the design at the cost of adding analysis blocks to a design. In many systems, instruments provide the most straightforward means for embedded trace or to directly configure, take direct control, or inject stimulus into a subsystem, as needed to resolve system level bugs.

A typical debug flow consists of several diverse tasks, both independent and interdependent, required to achieve a level of comfort in verifying an in-silicon product. With many devices consisting of both processor and fixed IP, along with related software and firmware, the verification concern is not only *operating as designed*, but also *performing as required* in its natural environment. For many products, this may include being exercised and verified in operational scenarios that were not foreseen or feasible to include during the pre-silicon verification cycle.

In recent years, in-system debug has taken on a specialization of its own, referred to in different contexts as on-chip instrumentation, design for debug (DfD), and the like. A flow of debug and analysis tasks that can be provided using instrumentation consists of several diverse independent and interdependent activities required to address different aspects of verifying an in-silicon product. DfD methodologies are still emerging areas of investigation. DfD differs from DFT and related approaches in the level of customization required to support specific debug requirements of an architecture or system (Fig. 1.8).

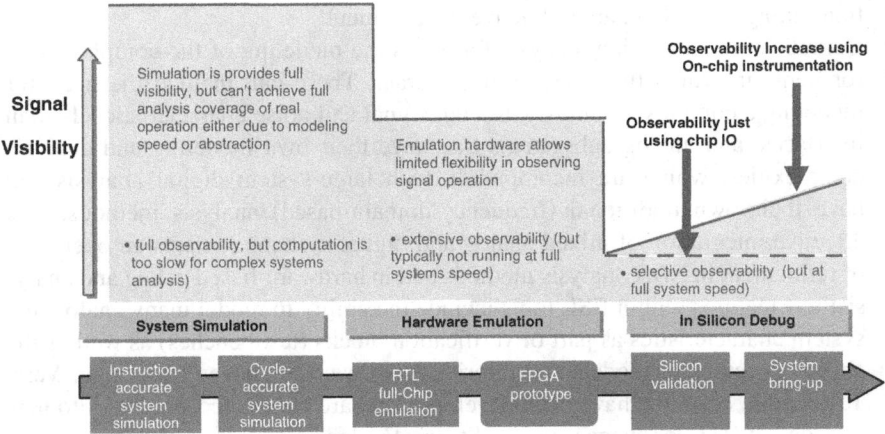

Fig. 1.7 Observability during design flow stages

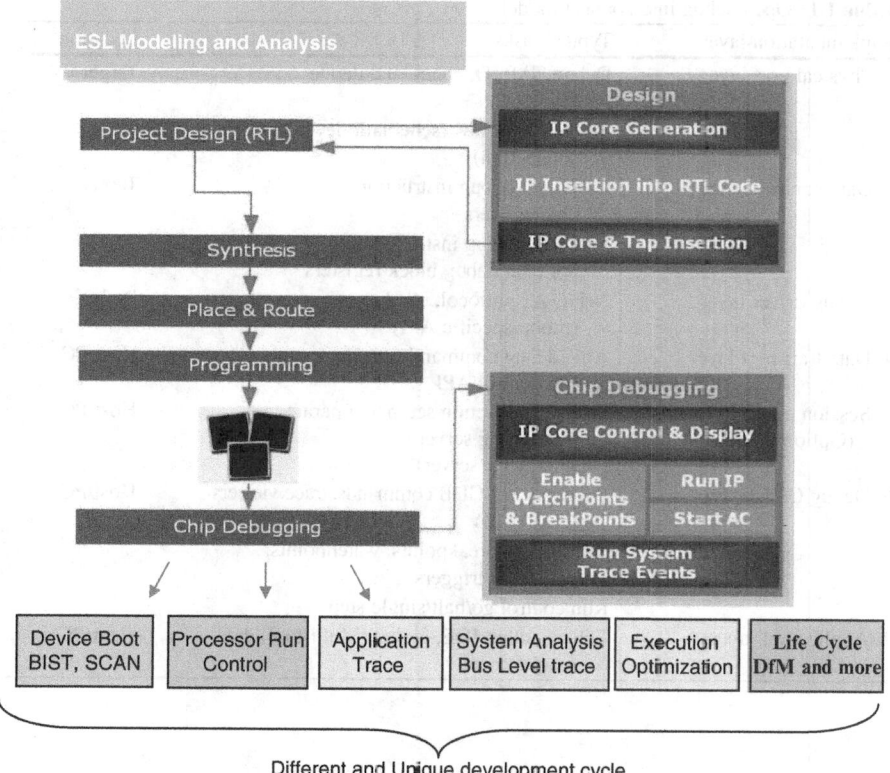

Fig. 1.8 Debug activities in the in-silicon verification

The availability of gates and on-chip resources of modern SoC allows for more innovative approaches to systems debug and embedded logic analysis by allowing dedicated debug subsystems to be created, with minimal or even negligible impact on the overall chip size. Dedicated debug subsystems would effectively extract and analyze signals and operations within and between deeply embedded processor subsystems of a complex design.

1.4 Instrumentation-Based Debug Infrastructure

A debug reference model shows a properly designed debug environment; different debug systems may be created in a modular fashion. Although a majority of the layers are software implemented in the debugger host, the two key instrumentation layers (1 and 2) in hardware address different instrument blocks that operate largely independently. Layer 1 I defines the port interface and its logic. Layer 2 defines the instrument function and operation. This separation allows the configuration of instrument related registers and decode of debug instructions can be treated largely independent of the physical layer, be it JTAG or any other interface (Table 1.1).

Table 1.1 Open debug interconnect model

Implementation layer	Typical tasks	Location
1. Physical port layer	Debug TAP IO, chain, and debug block wires	Target
	Debug TAP FSM (schematic-level connection)	
2. Data control layer	Low-level debug instructions and registers	Target
	Extended debug instructions, optional debug block registers	
3. Debug driver layer	Debugger protocol, clocking (probe-specific API)	Probe
4. Data transport layer	APIs debug command sets, run control API	Host PC
5. Session control layer (Optional)	Device connection setup and parameters	Host/PC
	Remote debug server (e.g., GDBserver)	
6. Debug GUI layer	Debugger UI, GDB commands, trace viewers (e.g., VCD)	Host/PC
	Set/observe breakpoints, watchpoints, and event triggers	
	Run control go/halt/single step	
7. Application layer (Optional)	Eclipse, other IDE, global (Multi-tool) data management	Host/PC

The integration of deeply embedded memory and embedded buses, along with limited IO for such embedded subsystems available for test purposes, limits the visibility of the embedded processors in SoC operation and dataflow.

In formal testability terms, multicore embedded systems present an asymmetrical functional test problem. Their controllability is high, because the systems are dominated by programmable processor cores. The observability is low, however, in terms of both critical signals that are directly available and the amount of embedded logic and internal signals as a ratio of the available IO in which to observe them. The addition of dedicated resources and structures that support functional analysis is needed to increase system observability. This requires a hierarchical focus to the issue of system analysis, starting at the individual core level of debug instrumentation and resources and increasing to a more system-centric diagnostic capability to facilitate increased observability. Although embedded debug instrumentation approaches are becoming increasingly common at the core level, system-level diagnostics and analysis at the multicore level has been a largely underaddressed and unresolved area of focus in complex embedded systems.

These "deep encapsulations" of key system functions, along with higher internal bus speeds, make traditional debug techniques, such as emulators, so limiting that they have forced the evolution to new logic analysis and debug approaches such as on-chip instrumentation.

Based on the shortfalls in applying current debug approaches to complex SoCs, the debug of structured SoC and related single-chip systems containing many embedded processor cores requires new system-level instrumentation approaches. The integration and debug of multiple cores, combined with an increasing ratio of overall gates versus package IO, makes an increasingly dominant portion of a system design "deeply embedded," so that only a minimal amount of data is needed for analysis to be made available in real time at the chips pins. These deeply embedded systems introduce new analysis problems, due to the interaction and communications of multiple cores, in addition to the more traditional debug issues associated with single-processor systems. The multicore debug requirement implicit for SoC requires new capabilities that exceed what can be addressed by traditional in-circuit emulation and logic analyzer capabilities, and by JTAG and BDM resources used in single-processor architectures. Whereas a JTAG or BDM can provide a snapshot of a piece of the system, the dynamics and interaction of multiple processors require a more dynamic and robust means of providing diagnostic information necessary to the designer and integrator (Fig. 1.9).

On-chip instrumentation is implemented as an embedded block that provides external visibility and access of the "inner workings" of processor and system architectures. When properly implemented, it provides a real-time "peephole" into the operations of key internal blocks that cannot otherwise be accessed in sufficient granularity on a real-time basis. The real-time visibility and monitoring of key interfaces and buses are increasingly crucial to understanding the dynamics of the operation of system architectures. As a general rule, debug visibility becomes

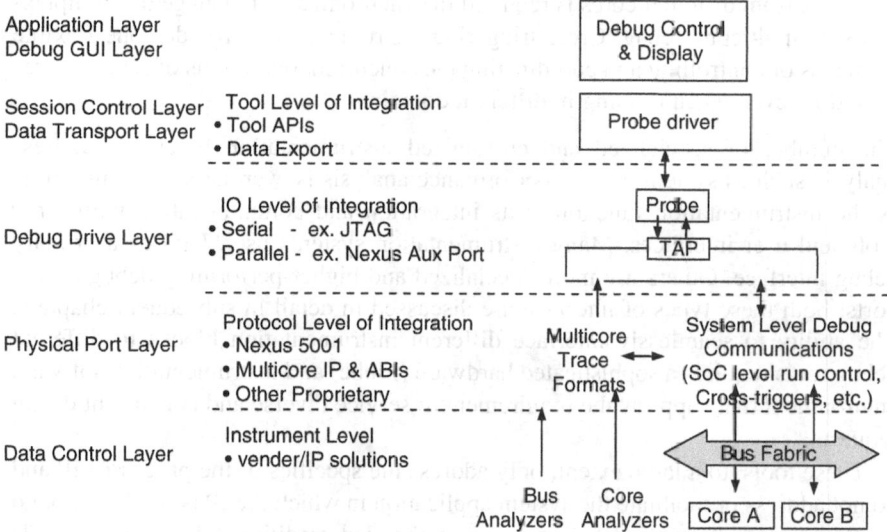

Fig. 1.9 Open debug model and components

increasingly problematic for highly integrated chips, which have extensive on-chip memory and caches, peripherals, and a range of on-chip buses. The key control and bus signals of interest in a deeply embedded system are often not externally addressable by the physical pins of the device and therefore are inaccessible to traditional instrumentation. This accessibility issue inhibits verifying silicon operation, introducing many hardware and software integration roadblocks, because the design team must address how traditional debug tools can be interfaced to work properly in SoC designs.

The value of instrumentation is, directly and indirectly, a function of several factors, including the instrument resources inserted on chip, the cost of the instrumented code and logic to the overall system, the overall applicability of the instrumentation, and the level of software and tool support available to make use of the instrumentation. In looking at the different types of on-chip instrumentation, they break out into roughly four major types of functions.

- Core Debug – most processor IP includes some debug blocks that simplify run control (e.g., go, halt, single step,) and optionally provide instruction and data trace. The core-level integrated debug blocks and debugger features can differ significantly from processor to processor.
- Logic Debug – providing more generic control and trace, logic debug IP essentially allows the embedding of a logic analyzer interface and part of a logic analyzer itself on the chip to provide visibility (and sometimes control) into the IP operation by allowing data capture of deeply embedded signals.
- Bus Debug – embedded bus fabrics provide data movement between cores and present additional challenges for system debug due to complex interactions of on-chip bus fabrics and the sheer amount of data transferred over bus channels.
- System Cross Triggering – for multicore systems, controlling and monitoring events from different cores is required to synchronize and manage the complexity of multicore debug. Cross-triggering instrumentation provides one flexible means of controlling and coordinating the concurrent operations of several cores and IP, even when running in different domains.

The number of specialized and customized instrumentation blocks to address analysis such as system or core performance analysis is even larger. As important as the instrumentation function is its integration and communication with other tools and user interfaces. Many instrumentation systems use JTAG as a primary debug interface. Others use more specialized and higher-performing debug access ports; both these types of interface are discussed in detail in subsequent chapters. The ability to seamlessly interface different instrumentation blocks to different debug tools requires a sophisticated hardware (probe) and instrumentation software environment that supports the requirement to service diverse and concurrent debug requests.

These tools, to a large extent, only address the specifics of the processor IP and do not address or facilitate the system application in which the IP is used. Although the processors become increasing deeply embedded, traditional development tools for system debug applications can not provide nonintrusive visibility into the highly integrated embedded processor. Applied to processor in-circuit emulators and their derivatives such as JTAG hardware debugging, the system must be placed in special debug modes or halted before being it can probe processor registers or read/write to the embedded memory. In many cases, this interruption of the steady-state performance of the system introduces (time) intrusive elements into the system operation that can complicate or invalidate the data or operations being observed. This problem grows proportionally to the ever-increasing frequency and complexity of high-performance embedded processors.

Chapter 2
On-Chip Instrumentation Components

In this chapter we examine a typical on-chip instrumentation environment and discuss some of the individual instruments used for system debug, their typical features, and their integration, both with each other and the systems being analyzed.

2.1 Trace and Event Triggering

Concepts of the tracing of data as it moves through the application or system are central to most other instrumentation capabilities. To address different debug requirements, instrumentation blocks must support different implementations of trace collection. Typical requirements include the ability to trace in cycle, branch, and timer modes. Cycle mode collects all bus cycles generated by the core(s). Branch mode collects all execution path changes, sometimes called branch trace messages. Timer trace mode records a frame with a timestamp each time an event is satisfied, providing basic performance analysis measurements.

Event recognition is widely used in conjunction with trace to capture information on events and operations in the SoC. Trace data values can be monitored and compared to provide real-time triggers to control event actions such as breakpoints and trace collection. Event recognizers can simultaneously look for bus address, data, and control values and be programmed to trigger on specific values or sequences such as address regions and data read or write cycle types. The event recognizers can control enable or disable of breakpoints and trace collection (Fig. 2.1).

Data tracing based on recognizable events opens doors to new capabilities in real-time SoC analysis. The data trace mode provides real-time information about the status and data of a system's internal signals, including, for example, analysis of cache performance and internal memory and data transfer operations that cannot otherwise effectively be extracted from a system. In-line or postprocessing trace information allows for analysis of data flow performance or measurement of system characteristics such as bus availability or cache hit/misses, which require long-term steady-state (measured over many cycles) system information. Additional detection of events in traced data allows the development environment to flag specific features in the trace data as it flows through the application.

N. Stollon, *On-Chip Instrumentation: Design and Debug for Systems on Chip*,
DOI 10.1007/978-1-4419-7563-8_2, © Springer Science+Business Media, LLC 2011

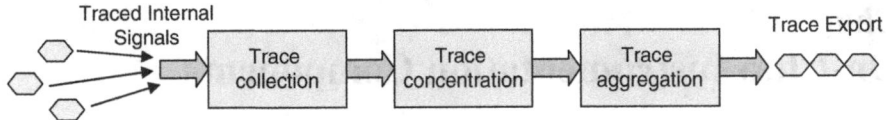

Fig. 2.1 On-chip trace formatting and export instrumentation

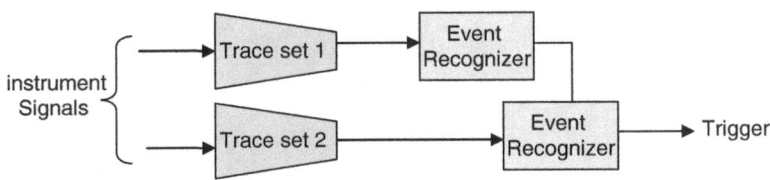

Fig. 2.2 Instrumented event recognition

As an example of a complex instrument event recognizer feature, four event recognizers can be combined in a 1–2 and 3–4 arrangement to produce two complex events. In this arrangement, the complex events can be configured so that first event of the event pair must be satisfied before the second event is enabled (Fig. 2.2).

2.2 External Interfaces for On-Chip Instrumentation

JTAG pin interfaces are the default interfaces of the most basic debug functions. Higher pin-out trace and probe ports are used with many on-chip instrumentation approaches. Even with these ports, however, the amount of debug information available can easily exceed the allocated debug interface of a SoC. To reduce the information being sent over the interface, approaches such as data compression increase the performance of the debug interface without significantly affecting system cost.

Obviously, the most useful approach to reducing the information from the debug port to the host development tool is to limit transmissions to new information and have inferred information derived by the development tools. For example, for required addresses to trace the instruction flow, it can be seen that not every instruction is required to construct an instruction trace. If the target processor does not have a change of flow, then the full address does not need to be transmitted. Only when a change of flow, such as an interrupt or branch, occurs would the system need to send the new beginning address. In addition, if the debugging session must be real time, then some information may be held in reserve. For instance, not all data values have to be visible at all times; only the data that the engineer is concerned with should be sent to the debug port during run time.

One of the major limiting factors on the use of instrumentation in SoC and multicore architectures is the ability to quickly export data as it is generated. On-chip instrumentation can address many of the operations associated with large amounts of on-chip debug, including triggering and performance monitoring. There is, however, almost always a need to be able to view the debug signals such as instruction/data trace from a processor, which means data must be exported off chip. The ability to transmit debug signals, most notably trace, is a hard limited function of two parameters:

1. The number of IO pins that can be dedicated to export of debug information at any given time.
2. The speed at which these signals can transmit the data.

This problem of exporting debug data is compounded for multiple-core SoC architectures, with monitoring of internal address, data, and control signals for each core, with the addition of inter-core and peripheral bus signals. One basic instrumentation approach is to rely on on-chip memory to buffer between traced data and the export bandwidth available. Trace buffer must consider the differences between data being generated on chip and the throughput of the debug interface. If buffers are modest in size, they can be overloaded by a large amount of trace data, as example, from multiple IP blocks or internal buses.

Despite the increasing number of IO signals available in leading-edge packages, system designers must limit the number of IO signals dedicated to trace and debug to reduce system cost (with packaging becoming an increasingly dominant factor in system cost). Most current approaches to increasing the IO bandwidth for debug rely on increasing the effective number of IO pins available (by multiplexing debug mode information into other system pins) and using higher-speed IO to increase the throughput of each pin. Each of these approaches to increase debug throughput has advantages and disadvantages. Increasing the effective pin count by statically multiplexing pins is a well-proven and low-risk approach. It does, however, involve coordination over the entire operation of the SoC, because pins that are dedicated to extended clock cycles to debug operations are unavailable for use in other modes of operation. To support SoC core and internal bus speeds, bigger pin bandwidth is increasingly required for instrumentation interfaces of a SoC.

2.3 Performance Analysis Using On-Chip Instrumentation

Customized instrumentation can integrate performance analysis of SoC architectures as part of a debug solution. Performance analysis (PA) is an all-encompassing term that refers to many types of measurements that provide information on how a particular core is being used, both in context of other parts of the system and with regard to specific algorithms. Integrating instruments to allow processor characterization, software performance, and system performance metrics provides valuable and concise information, which is more simply gathered locally to the processor because the lack of IO signal visibility in individual processor operations limits tracking of embedded processor performance. Performance metrics can be distorted or obscured by the layers of system buses, peripherals, and limited IO access between an embedded processor and the external test environment.

Some common types of tests that are desirable in processor and SoC system performance analysis are:

- Find and profile hot spots in execution.
- Be able to measure loop times.
- Trace function calls, returns, and interrupts and measure the performance of this code.
- Measure the duration of ISR (interrupt service routine) and other events.
- Track interrupts and measure the maximum interrupt latency.
- Track RTOS context switches, measure task duration, and measure OS events such as semaphore waits.
- Measure the cache hit/miss ratio.
- Measure on-chip and off-chip memory access use.
- Count the number of processor stalls caused by (slow) bus accesses.
- Measure bus use and which master-slave transactions are using the bus the most.
- Count the number of processor stalls in a section of code.
- Count the number of instructions executed between two points in a program.

2.4   On-Chip Logic and Bus Analysis

Instrumentation-based logic trace allows analysis of bus architectures and related nonprocessor IP. Logic analysis instrumentation typically consists of debug blocks that are integrated into synthesizable logic files (typically VHLD or Verilog).

The bus analyzer collects a history of on-chip bus activity and exports it through the JTAG interface. Bus signal information is connected to the data inputs. A triggering system user starts and stops collection of data to an on-chip trace RAM. When collection stops, the most recent activity remains in the trace memory, from which it is unloaded through JTAG and displayed. The bus trace configuration includes a timestamp, which is stored with the data; to provide synchronization and interval information, on-chip counters for performance measurements; of the frequency of system events, and JTAG-controlled registers that hold parameters for input and output triggering of control operations that allow captured bus signals to interact on-chip with other debug components in the system.

- Bus fields include address bus, data bus, and user extension field and can track a number of bus masters in the system. More than one bus layer may be supported in a single instance. For more trace capability, or trace over different clock domains, more than one bus navigator instance can be implemented in a single JTAG chain.
- The trigger state (started, active, stopped, stalled) is recorded in the trace buffer. A multistate trigger allows triggering on sequential events. For example, a configuration that recognizes bus cycle A followed by bus cycle B is:

```
if (event A and state 0) then goto state 1
if (event B and state 1) then trigger
```

- Timestamps are used to indicate the distance between recorded samples when collecting trace using qualifications such as trace-on/trace-off, or collecting filtered trace that matches an event definition. Because bus measurements may be large numbers of cycles, the timestamp is set up to cover a large time range.

Being able to trigger from instrument data allows for both dynamic interactions with the target system and improved capture of the information of interest. Analyzers nominally support multiple triggers with multiple states per trigger (Fig. 2.3). Trigger conditions can be created as application-specific combinations of three components:

- Raw or processed data (filtered or aligned) compared to logic or edge events on each signal.
- Counter or times values matching a preprogrammed value.
- Trigger state (what trigger-related operations have occurred previously?).

When a trigger condition is satisfied, one or more actions can be taken, such as to mark the trigger frame, turn trace on or off, record a single frame, turn the counter on or off, increment or clear the counter, assert the external trigger out, or change the trigger state. The flexibility of this system under a wide variety of conditions and actions can improve visibility and monitor and tune system performance based on a range of operational parameters.

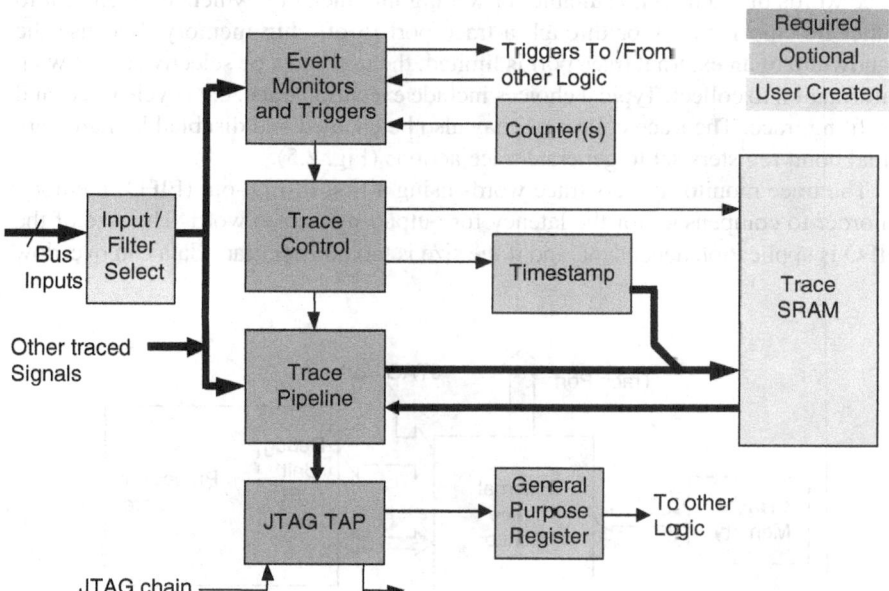

Fig. 2.3 Bus trace instrumentation block diagram

2.5 On-Chip Instrumentation Examples

In this section, we present several examples that illustrate the instrument features just discussed.

2.5.1 Trace Monitoring and Interfaces

Embedded processor instrumentation addresses embedded processor debugging and system validation features such as run control, trace history, memory and register visibility, and complex breakpoints.

The external trace monitor is an instrumentation block integrated into and supporting of processor core monitoring. Trace monitoring allows capture of both execution history and other real-time information from the core and allows either on-chip or off-chip trace storage. Trace monitors can also be configured to collect profiling data for performance analysis. The specific instance of the instrument interfaces a debug unit interface for a processor architecture that provides debug functions such as start/stop execution, single-step, breakpoints, and register/memory access (Fig. 2.4).

The trace monitor allows trace history to be captured in several modes (instruction and/or data full or compressed, etc.), depending on the available bandwidth and information desired. The block combines trace messages of various lengths into trace words of fixed width suitable for writing into memory, which are then sent to either on-chip memory or through a trace port to off-chip memory. Because the bandwidth of an external trace port is limited, the user must be selective about what information to collect. Typical choices include execution trace, data cycle trace, and profiling trace. The trace collection may also be enabled and disabled by hardware breakpoint registers set to generate trace actions (Fig. 2.5).

The trace monitor buffers trace words using a first-in-first-out (FIFO) memory, in order to compensate for the latency for outputting a trace word. The size of the FIFO is application dependent, and if the size is too modest, trace data can overflow

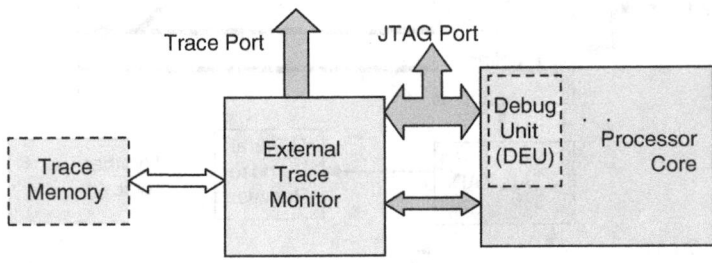

Fig. 2.4 Processor trace

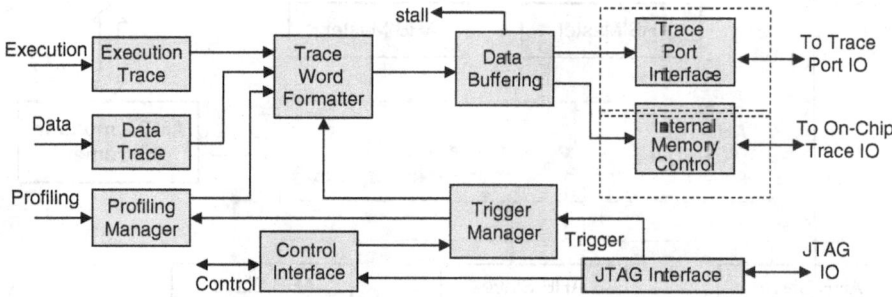

Fig. 2.5 Trace monitor block architecture

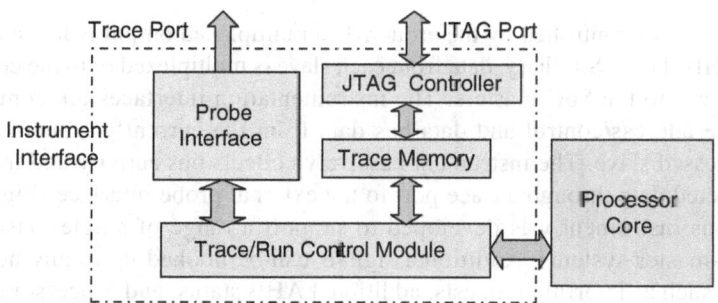

Fig. 2.6 A processor-instrument interface

and become corrupted. The trace monitor control logic allows requests that the processor pipeline be stalled so that no trace information is lost (Fig. 2.6).

The trace monitor allows internal or external trace memory. When data is available from the data buffer, it is written to the internal memory. For external trace, the off-chip trace port logic multiplexes the trace words from the data buffer onto the trace port pins. As in the previous examples, control of the instrument block is handled by a JTAG interface and can be configured for on-chip or off-chip trace storage.

2.5.2 Bus Logic Monitoring

With increasing core density and interconnection of blocks in modern SoC design, monitoring internal bus operations is an important capability to debug the entire SoC. OCP and AMBA AHB are leading on-chip buses in use by many SoC design architectures to communicate between cores. On-chip instrumentation applied to the AHB captures bus activity and allows system diagnostics in real time.

In this case, the instrument connects to the AHB address bus, data buses, and various control signals at the bus multiplexed outputs. In AHB, signals are driven from each master and multiplexed onto a common address/data/control bus

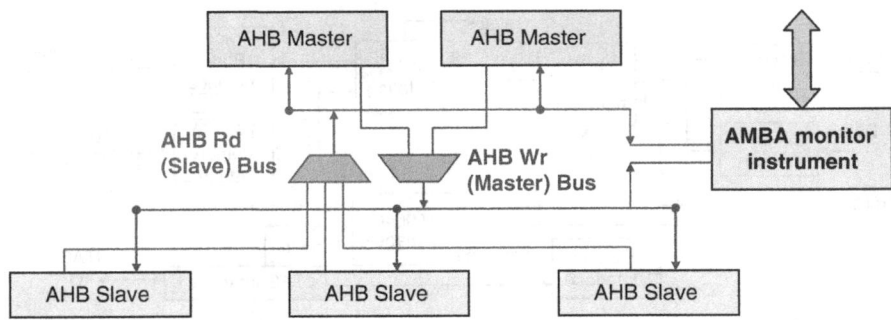

Fig. 2.7 Instrument interface into an AMBA bus

by a multiplexer controlled by the arbiter. The multiplexed output is fanned out to all the AHB slaves. Similarly, data from each slave is multiplexed onto the common bus and sent to the SoC masters. The instrumentation interfaces are configured to receive address/control and data bus data from the currently granted master and addressed slave. The instrument passively collects bus activity and transfers the collected data through a trace port to the external probe interface (Fig. 2.7).

The bus instrument was developed to support a range of single-master and multiple-master systems. Additional signals can be hooked up to any nodes in the SoC, such as interrupt requests, additional AHB status, and processor control signals. The additional signals can also be used to compare and recognize specific on-chip activity outside the AHB bus, and then are transmitted to the probe for triggering purposes. As an example of real-time processing for debug that the instrument enables, the bus monitor allows probing of data in different modes. In the AHB case, data can be probed in two modes. Bus-cycle mode captures all address/control and data signals exactly as they occur per clock on the bus. Bus-transfer mode reduces the delays and latencies between address and data cycles on the bus, by aligning to the same clock cycle, operations that occur in different cycles. This reduces the number of trace cycles that are stored and allows for efficient combination of address-data-control event triggering for trace and monitoring operations. Bus transfer mode is especially effective for bus read operations in which the master transfer operation providing addresses and control and slave response providing data back to the master may be separated by a large number clock cycles of the bus waiting for the operation to complete. As an additional example of trace in-line processing, the trace hardware can be configured to filter out idle, busy, and not-ready cycles where no active data is being transferred. This allows each trace frame to record the critical AHB signals along with additional user-selectable signals and a timestamp to aid in performance analysis.

The host software for AMBA monitoring provides a good example of a specialized debug interface to support bus operations. Bus values can be viewed either numerically or symbolically. The symbolic representation increases the visibility and comprehension of complex bus operation (Fig. 2.8).

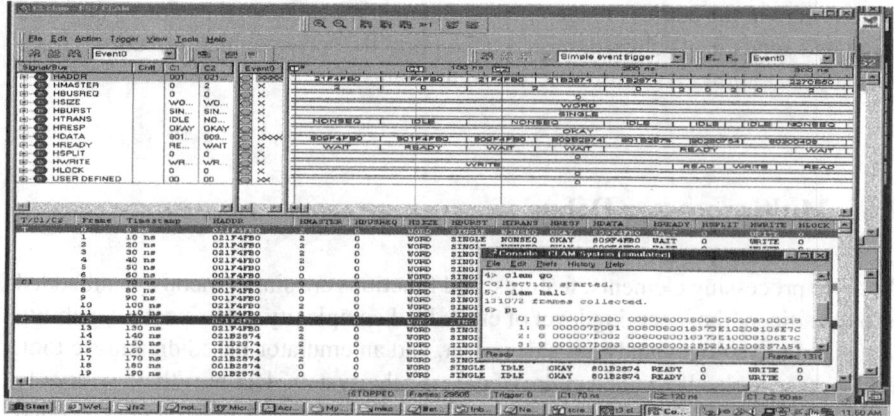

Fig. 2.8 A bus analyzer display

2.5.3 Real-Time Data Exchange

Real-time data exchange is the ability to "exchange commands and data with the application while it executes." This approach to "dynamic instrumentation," called "dynamic variable adjustments" or "dynamic data collection," was introduced by Texas Instruments and is becoming widely utilized. Dynamic data collection refers to the ability to capture specific address ranges of data from the SoC target and present them to the user on the host machine. The data can be "pulled" periodically by instrumentation or on-demand by the user using the JTAG and/or trace port. Pulled data-exchange methods of implementation can include a JTAG command that suspends the processor, reads a range of data values from the target, and passes them to the host via the probe interface.

Debug data can also be "pushed" from the target based on instrumented code to output variables or arrays periodically (i.e., timer interrupt) or from executing a specific location in code – such as when a variable is updated. Pushed data exchange can be implemented based on instructions in the target code, such that a range of data will be copied from memory to the instrumentation trace port. The hardware core and instrumentation block provide an instruction that can write memory to the trace port or a DMA channel configuration that can do a range transfer from memory to trace port. The data format can function as burst mode – first the start and end addresses are sent out (or start address and length,) followed by the data. If the trace port is not available, a breakpoint can be placed in the code and the run control unit fetches the string buffer via JTAG reads.

A third technique is the use of "shadow memory" – an external RAM that is interfaced to hold the same image of values as in the processor's main memory (or cache). Shadow-memory techniques include zero-overhead methods in which the instrumentation is set up with a range of addresses to shadow. When a read or write occurs in the address range, the instrumentation captures the address and data value and sends it off-chip through a trace port. To a probe which includes a trace memory that allows a real time access to the data.

2.6   Multiprocessor Debug

As more processing elements, features, and functions are simultaneously embedded into the silicon, the emerging level of embedded complexity outstrips the capability of a stand-alone logic analyzer, a debugger, and an emulator-based diagnostic tool. While these tools allow the capture of data off the system data bus, they work only as long as every access (read or write) occurs over the external data bus. For embedded processors and buses with no direct external access, this points to a growing gap in effectively being able to provide the necessary controllability and, in particular, the visibility of the internal operations of a complex system.

The need for improved methods of observing and analyzing embedded processor and SoC operation has increased at a pace at least proportional to the explosive growth in SoC designs and new IP cores. This forces the analysis side of the SoC world into a constant process of catching-up to the designer's ability to add cores and integrate new resources on chip. With an ever-shortening development cycle, and often several generations of products being produced in parallel or rapid succession, standardized embedded tools and capabilities that enable quick analysis and debug of the embedded IP are a critical factor in keeping SoC verification a manageable part of the process.

Before we can implement an on-chip debug system suitable for multicore systems we have to ask the user requirements.

1. Each core and bus must be observable. We must be able to see or reconstruct the program flow of each single core independently as well as of the data flow on the system buses. Also important are signals allowing conclusions to be drawn about power modes, bus accessing modes, and others.
2. It is crucial for system analysis to recognize events that arise from interactions between the cores and buses. A single core on its own is no longer of interest. Rather, events coming from several cores have to be considered. To minimize this challenge, cross triggers must be used, which combine events from different sources and make them available systemwide.
3. The interactions between all SoC components during debug become more complex as more components are involved. A debug system with complex cross triggering is hard for the user to manage. The debugger as a user interface for the complex debug hardware must support the user in its work finding the mistakes or performance bottlenecks. It has to hide the complexity that comes with multicore debugging. We must not forget the user's task is to cope not with the debug hardware itself but with the faulty system.

On-chip instruments (and simulation) play an important part of SoC development and verification flows, providing the ability to analyze what is happening on the hardware itself, during both prototyping and system-level verification stages, and increasingly on the final products themselves. The problem in analyzing information like embedded buses in hardware in many cases hinges on a problem of visibility: *it is difficult to fix what you cannot see.* This visibility problem for the embedded SoC is more complex than can be addressed adequately by traditional on-chip test methods such as traditional JTAG scan, for several reasons:

- Bus operations are multicycle, with signals in a bus cycle becoming active at different times, requiring sequential tracing rather than as a single-cycle snapshot that scans typically provide.
- Bus operation problems are interrelated with the operations of at least two communicating blocks (a processor and memory peripheral, for example). Traditional debug methods, such as halting part of a system for testing, can introduce changes and new variables that interfere with the test scenario and process.
- If problems are intermittent or sparse, then trace operations need to operate in a triggered mode, so information for a given range of bus cycles of interest is captured in real time.

The problem is, to a large extent, a multicore extension of embedded processor analysis, with run control, instruction execution, and data trace as integral parts of processor support. For larger systems with multiple cores, the problem extends beyond processor execution to understanding system operation and communications (Fig. 2.9).

In formal terms, multicore embedded systems present an asymmetric functional test problem. Their controllability is high, because the systems are dominated by programmable processor cores. The observability is low, however, in terms of both the critical signals that are directly available and the amount of embedded logic and internal signals as a ratio of the available IO in which to observe them. Adding dedicated resources and structures that support functional analysis is necessary to increase system observability. This requires a hierarchical focus to the issue of system analysis, starting at the individual core level of debug instrumentation and resources and increasing to a more system-centric diagnostic capability to facilitate increased observability. While embedded debug instrumentation approaches are becoming increasingly common at the core level, system-level diagnostics and analysis at the multicore level have historically been a largely underaddressed area in complex embedded systems.

Single-core approaches for debug and trace often fall short when used with multiple cores and processor interfacing with complex application-specific IP. Increasingly, SoCs integrate multiple types of cores, either for DSP or other specialized processors or for other complex application-specific IP operations for a myriad of functions. These cores may be running asynchronously or with variable or indirect communications with each other, which makes debugging over multicore difficult to correlate. Complicating multicore debug issues further, in many cases,

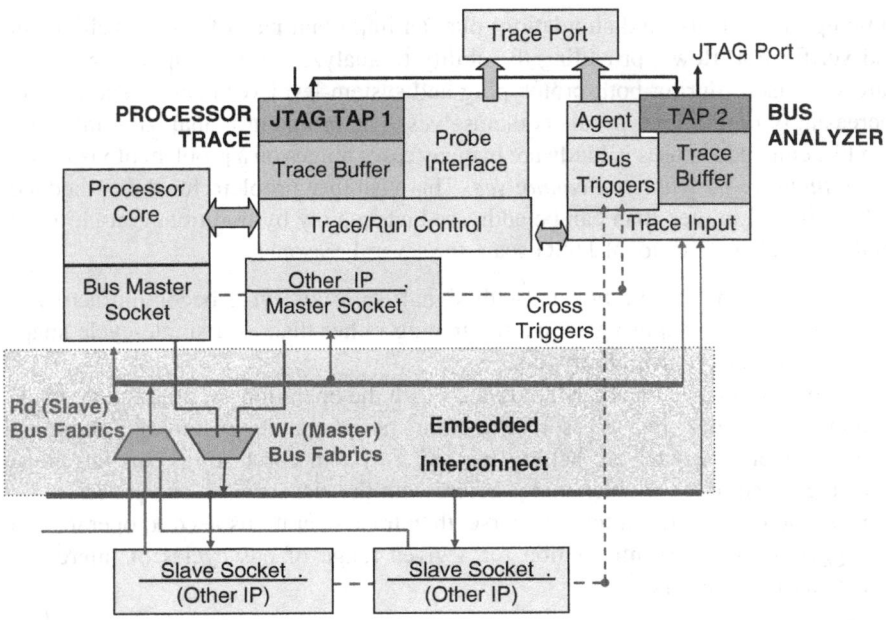

Fig. 2.9 SoC processor and bus trace instrumentsssss

different IP blocks come from a variety of vendors and have different compile and debug environments or levels of debug features. Tasks such as processor interfacing, interprocessor communications, run-time execution and coordination, and data presentation place a significant overhead on the debug requirements for heterogeneous and multicore chips. In these architectures, a range of the instrumentation block(s) must be customized to support the specific verification and debug requirements of both processors on a stand-alone basis and in a multiprocessing configuration. Among the basic requirements, instrumentation blocks must be diverse enough to effectively communicate debug data with their respective cores and have a sufficient common interface to coordinate all their activities. For example, synchronization of all processors in SoC is required in starting and stopping their operations.

Instrumentation solutions for on-chip buses provide a valuable resource for observation in multiprocessor debug. For systems that have multiple processors communicating over a standard bus such as AMBA AHB, access to information such as which processor owns the on-chip bus can provide valuable context as to what the relative communication and stages of processor execution are. With increasingly complex bus architectures being introduced, it is generally agreed that future generations of multiprocessor debug will rely on more extensive tracing and triggering of bus operations to address interprocessor communication issues in conjunction with more specific point solutions for processor-specific analysis.

Looking ahead to more complex systems, instrumentation must have sufficient "embedded intelligence" to interpret information passing between cores, determine what is needed to be extracted for debug, and perform other task-aware debug for on-chip RTOS or network protocol analysis. Equally challenging is presentation of all the diverse debug information in a coherent, understandable way. As in many areas of complex SoC design, new classes of instrumentation are needed to address diverse debug and analysis requirements of emerging architecture.

Integrating instrumentation into design hardware enables on-chip debug capabilities by providing not only visibility but also control features such as breakpoints for developing and integrating SoC application code. On-chip instrumentation and debug are critical resources to aid in both processor function and performance assessment and effectively evaluating silicon prototypes (for example, programmable logic implementations) and first silicon debug and validation.

There are a wide range of approaches taken for embedded processor debug, several of which are discussed in later chapters. There is no magic bullet instrumentation approach, but rather a number of commonly required capabilities needed to provide a robust debug solution. On-chip instrumentation enables a range of widely used best practices in debugging and interfacing embedded information, including data tracing, triggering run control that has proven analysis benefits. Implementing an instrumented interface on SoC designs offers distinct advantages in efficiently implementing run control, real-time instruction and data trace information, RTOS support, memory subsystem, breakpoints, and watch-points, to name just a few.

An instrumentation implementation that is scalable and configurable to map to a range of instrumentation requirements on the SoC allows support of user-definable general-purpose or application-specific features. Instrumentation hardware should be a synthesizable solution, both to facilitate integration into a range of target platforms and to load instrumentation into hardware emulators to provide a synchronized method of loading and debugging code and functionality in a pre-silicon environment. Synthesizable instrumentation solutions also allow their integration into high-end FPGA parts. In many cases, programmable devices incorporating instruments through their system interfaces are ideal for pre-silicon verification.

One of the most important features of instrumentation capability is support of collection and streaming of data off-chip to a logic analyzer or other trace postprocessing environments, which integrates trace processing along with a low overhead control interface.

Support for complex event recognition and triggering capabilities is also required to provide a robust level of control and monitoring of operations. Complex address and data triggers, coupled with bus trace, can be used for a range of operations from multiprocessor synchronization to debugging device drivers. Having a source-level debug GUI coupled with the instrumentation complex triggers may rapidly uncover execution errors and problems such as improperly defined variables. By coupling timestamps to trace data, complex triggers can be used to provide a range of performance analysis information.

The ability to interpret debug information is essential. A documented API allows fast, efficient porting of instrumentation to customer-specific GUIs. Scripting of validation and manufacturing tests is a useful means for efficiently leveraging embedded instrumentation. Host debugger environments for an instrument solution benefit from command-line interfaces that allow effective script file usage.

Instrumentation solutions for processors should support complete integration into a source-level debug interface to provide access to disassembly information needed to understand the context of application-related problems.

Instrumentation extensions can be customized in a range of areas for the system to debug application-specific IP. Their value in providing otherwise unavailable visibility in a range of internal system characteristics, including code coverage, RTOS task analysis, and protocol analysis, will only increase with larger, more complex, and increasingly deeply embedded next-generation architectures.

Chapter 3
JTAG Use in Debug

IEEE Specification 1149.1, more generally known as JTAG (Joint Test Action Group), was originally developed as a test architecture with a standard serial interface to an on-chip test access port (TAP) to permit snapshot sampling of individual pin signals. It is, however, generic and flexible enough to also be used to load registers and drive specific output signals, which makes it capable of serving as a debug interface in a variety of scenarios. JTAG is probably the most widely used debug interface, as the JTAG TAP is found in most digital internal circuits (ICs) and JTAG debug interfaces are defined for most processors, FPGAs, and other commercial parts.

Most currently available cores and processor-based devices provide some form of JTAG interface for run control and debug functions. JTAG can be used to support trace and performance analysis instruments such as the ARM's ETM and Coresight, and MIPS EJTAG + TCB from MIPS; which allow integrated trace of both the processor and its system interfaces for more extensive system debug.

The biggest problem with on-chip debugging is the lack of a consistent set of capabilities and single communications interface across processor architectures. Using JTAG as a "debug port" has become perhaps the most widely used instrumentation interface, providing the stepping stone between traditional processor emulation and more SoC-friendly approaches to debug. JTAG was originally developed as a means of doing full chip testing and allowing serial testing of all the pin connections of a chip and its interconnections to other chips on the circuit board. Given that more than 200 major electronics manufacturers have adapted the JTAG standard since its release in 1990, JTAG is found in virtually all modern digital ICs. An 1149.1 JTAG TAP is a four- or six-pin interface that has both serial an parallel signals. Data is transferred between different TAG or with an external probe over serial test data input (TDI) and output (TDO) pins, and system-wide Test Clock (TCK). Test Mode Select (TMS) control, and (optional) JTAG test reset (TRST) are common to all TAPS. A sixth RTCK signal is not in the current standard but is widely used for debug related communications and is discussed later in this chapter. JTAG standard instruction operations define board- level IO testing via a Boundary Scan Architecture (BSA) that is based on chained scan registers that may be controlled through the JTAG Interface. This JTAG boundary scan architecture is primarily a connectivity test construction and is of less interest for debug. The JTAG specification allows alternative logic blocks to be connected to a TAP and the creations of user defined instructions.

N. Stollon, *On-Chip Instrumentation: Design and Debug for Systems on Chip*,
DOI 10.1007/978-1-4419-7563-8_3, © Springer Science+Business Media, LLC 2011

Most of the operations we will discuss use alternate logic blocks, that sits along side the BSA, and which defines instrumentation operations, Similarly, most of the debug related instructions we focus on are non-standard user defined instructions.

3.1 JTAG Pins

IEEE 1149.1 requires a minimum of four signals to support the TAP. TRST and RTCK are optional. Figs. 3.1 and 3.2.

Table 3.1 JTAG IO signals

JTAG I/O	Description of JTAG (IEEE 1149.1) Pins
TDI	*Test data input allows serial movement of data into the JTAG port.* Used to transfer instructions and data serially into the device. TDI is sampled on the rising edge of TCK and has an internal pull-up resistor
TDO	*Test data output allows serial movement of data out of the JTAG port.* Used to transfer data out of the device serially. TDO changes on the falling edge of TCK
TCK	*Test clock is an input pin that provides the clock for the JTAG port.* Used to sample the TMS signal, to strobe data and instructions into the device, and to strobe data out of the device
TMS	*Test mode select input provides access to the JTAG TAP state machine.* Used to change the TAP controller state machine to the next processing state. TMS is sampled on the rising edge of TCK and has an internal pull-up resistor.
TRST	*Test reset input optionally provides for asynchronous initialization of the JTAG IEEE 1149.1 controller.* Asserted low to reset the TAP circuitry to a known initial state. TRST is asynchronous to TCK and has an internal pull-up resistor
RTCK	*Return clock output can be used to accelerate data access through the JTAG port.* RTCK is not part of the 1149.1 standard

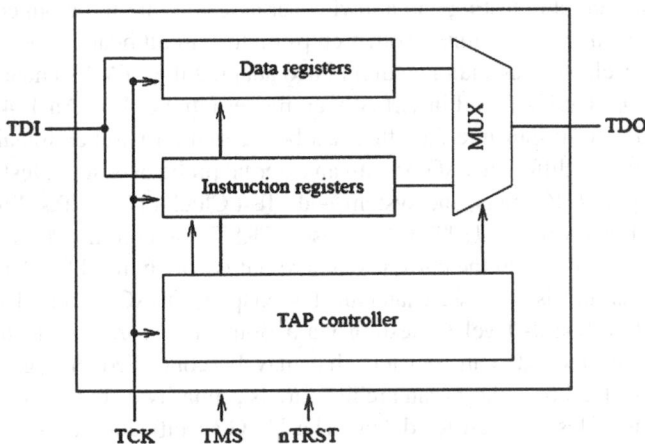

Fig. 3.1 A JTAG TAP

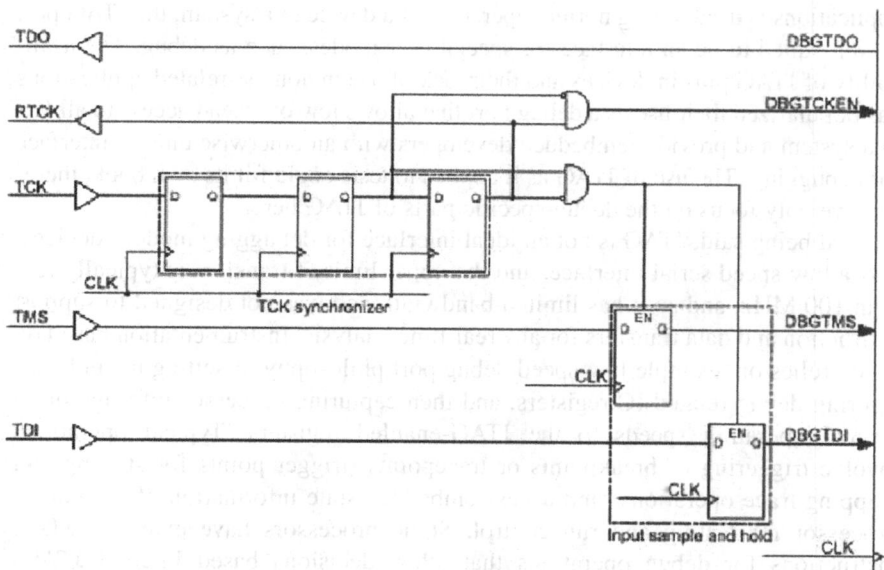

Fig. 3.2 JTAG interface using RTCK

Write operations pass data into the registers from TDI. Read operations pass data out of the registers through TDO.

The JTAG port itself has evolved in some circumstances to better support debug. One example of this is the introduction of return clock; RTCK is the return test clock signal from the target JTAG port to the JTAG interface unit. Some targets need to synchronize the JTAG port to internal clocks. To help meet this requirement, RTCK, which is a returned (and re-timed) TCK, can be used to dynamically control the TCK rate. RTCK is a synchronized logic clock consisting of a delayed version of TCK that is synchronized to an on-chip system clock to improve JTAG performance in trace operations. RTCK was originally developed by ARM but is now supported in general for many other processors' JTAG instrumentation.

Also widely associated with JTAG is a VTRef signal intended to supply a logic-level reference voltage to allow debug equipment to adapt to the signaling levels of the target board. VTRef does not supply operating current to the debug equipment.

Few vendors have standardized JTAG instruments outside of the physical port pin out and required test-related instructions for the physical connection. As a result, JTAG capabilities vary widely because on-chip debug was never addressed within the IEEE 1149.1 standard. The JTAG architecture defines signals and mandatory on-chip logic (16-stateTAP controller, instruction tegister, bypass register and boundary-scan register) that is also used by instrumentation. Chips may be daisy-chained together, connecting all registered I/O pins and buffers into a scan chain, where values may be read or written. For larger devices, this scan chain may be hundreds of elements long.

The JTAG port is a dedicated interface standardized for JTAG access. Additionally, it can be used for internal chip tests. Because neither of these applications is used during normal operation of a device in a system, the JTAG port is well suited to be an interface for special user modes, such as debug. The availability of JTAG pins in devices and their lack of use in nontest-related applications has popularized their use as a debug port that allows low overhead access to all ICs in a system and provides embedded developers with an otherwise unused interface for debugging. The use of JTAG as it applies to tests could fill its own book, therefore we only focus on the debug-specific parts of JTAG here.

That being said, JTAG is not an ideal interface for debugging modern devices. It is a low-speed serial interface, and the upper limit of transfers is typically less than 100 MHz, and so it has limited bandwidth and was not designed to support instruction and data transfers for any real-time analysis. Instrumentation based on JTAG relies on a simple low-speed debug port philosophy of setting up and configuring debug-related IC registers, and then capturing processor information at normal operating speeds to the JTAG-enabled registers. Typical operations involve triggering of breakpoints or tracepoints (trigger points for starting and stopping trace operations) and access embedded state information of the microprocessor for testing and run control. Some processors have extended JTAG instructions for debug operations that allow decisions based internal JTAG-enabled registers to change the processor state to debug intensive modes of operation for capturing system information. While in debug mode, the processor instrumentation can examine and modify the internal and external states of a system's registers, memory, and I/O space. A rich infrastructure of tool environments and standardized debug schemes has been built on this foundation to provide JTAG debug of both embedded processors and other parts of an embedded system.

The key to using JTAG as a debug port is the standard's provisions for user-defined instructions and data register sizes. Virtually all of the debug instructions and capabilities we discuss utilize this user-defined instruction capability. JTAG defines a small variety of standard test instructions with a low-overhead (and low-bandwidth) serial data access interface. Whereas JTAG defines standard test operations (scan in, scan out, bypass, etc.) that use defined registers, the JTAG standard allows for a much more diverse number of user-defined instructions that can be added.

Because JTAG does not define a fixed register size, the number of available registers that can be selected using different user-defined instructions can be much larger than the number of JTAG-defined registers for test purposes; the size of data registers associated with different instructions is also user-defined. Consequently use of JTAG for debug is essentially done on an ad hoc basis using the JTAG ability to access these user-defined instructions via a standardized and simple state machine in the TAP that serially scans instructions in and data in and out of the registers. These debug-mode instructions and their associated instruction and data registers must be designed into the JTAG block and must have a probe and other software infrastructure to support operations. Several locally standardized sets of debug mode instructions have been adopted for use with the JTAG TAP, especially with regard to processor debug operations. These include vendor-specific ARM ETM and MIPS EJTAG for the different processor families, as well as other standards such as IEEE 4001 (Nexus), IEEE 1149.7 (CJTAG/AJTAG), and IEEE P1687 (IJTAG) (all of which are discussed in later chapters). Each of these which define a JTAG-compliant protocol that supports additional debug functions, while adhering to the 1149.1 standard. All have support from a variety of third parties that supply probes and software tools that support these different JTAG variants. Of course, any chip developer may define and support its own debug instruction set by defining it in the JTAG TAP and on-chip instruments communicating with the TAP, and providing or developing its own infrastructure.

3.2   Test Access Port

The TAP is the external interface for the internal test circuitry specified by IEEE 1149.1. It consists of the following:

- 4 to 6 dedicated signal pins.
- A 16-state TAP controller.
- An instruction register.
- At least 3 data registers:

    - A bypass register (BR).
    - A device ID register (IDR).
    - A boundary-scan register (BSR)

The three defined data registers have defined purposes for debug operations, but there also must be at least one debug register. Using JTAG as a debug architecture requires debug-related registers connected between the JTAG controller and the internal device circuitry. Unlike a BSR, there is no pre-defined relation between the different debug registers. When selected by the appropriate TAP controller instruction, one or more debug registers becomes a serial scan path between a TDI pin and a TDO pin. During normal operation, the registers are static or in read mode to collect data. However, when the system enters debug mode, data loaded into the control registers is propagated and data from the system is captured for triggering and related functions. Data registers can be read or written by shifting in values from the JTAG TAP and applying them to an active register. Registered on-chip data can be exported through the JTAG TAP; for observation and analysis.

The JTAG controller state machine is the heart of the JTAG operations. It is also referred to as the test access port controller. The TAP controller is a 16-state machine that manages control of the JTAG environment to perform the instructions and data transfers between the on-chip registers and an external debug host. All state transitions occur on a positive TCK clock edge and are controlled by the TMS pin after reset (TRST). Initially, the state machine is in the *Testlogic-reset* state. With TMS low and a positive edge on the TCK, it is brought to the *Run test/idle* state. All further state transitions are done in a similar manner.

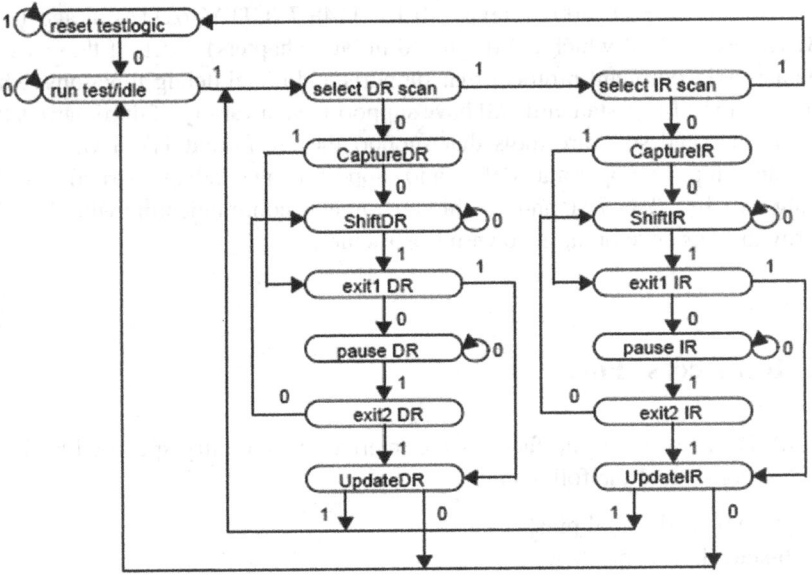

Fig. 3.3 A 16-state TAP controller state machine

Figure 3.3 shows the 16-state TAP controller state machine. The state machine performs three basic actions:

- Do nothing and wait for debug operation to be initiated by an external debugger host in the Testlogic-reset or Run test/idle state.
- Load a new instruction in an instruction register (IR) scan cycle.
- Load new data into a selected data register in a data register (DR) scan cycle.

The JTAG state machine has two parallel control paths. One is for the JTAG instruction register using the IR path, and the other is for the (selected) data register using the DR path. The instruction register directly or indirectly selects the register(s) for the next data operation. The IR and DR paths are identical to flow; the differences are in the registers that are being accessed. The corresponding IR and DR states are as follows:

- Select-IR-scan, select-DR-scan: initiate an instruction or DR access sequence.
- Capture-IR, capture-DR: load IR or DR in parallel.
- Shift-IR, Shift-DR: load data register by shifting data through the instruction register or the selected data register that is connected between the device's TDI-TDO.
- Exit1-IR,exit1-DR: finish phase-1 shifting of instruction or data.
- Pause-IR, Pause-DR: temporarily hold the access operation (allow the master to reload data).
- Exit2-IR, exit2-DR: finish phase-2 shifting of instruction or data.

- Update-IR, update-DR: parallel load of registers synchronizing the instruction
  or selected data register or instrument that is active under the current selected
  instruction.

In operation, the TAP changes state based on the level of TMS. Transitions from one
state to another occur on the rising edge of the TCK. Instructions and data are transferred through TDI, which is sampled on the rising edge of the TCK, and data is
transferred out through TDO, which changes on the falling edge of the TCK. This
sampling technique prevents the development of a race condition in the TAP. The
main state diagram consists of six steady states: Testlogic-reset, Run test/idle,
Shift-DR, Pause-DR, Shift-IR, and Pause-IR. A unique feature of this protocol is that
only one steady state exists for the condition when TMS is set high: the Testlogic-
reset state. This means that a reset of the test logic can be achieved within five TCKs
or fewer by setting the TMS input at a high level/high enough/sufficiently high/ etc.

A JTAG implementation consists of an instruction register and one or more data
registers, one of which is selected at any given time based on the contents of the
instruction register. The JTAG port master writes an instruction into the IR that
either performs an action or selects a particular data register, or both. The action
occurs when the TAP state machine passes through the Update_IR state. Status
information is returned to the external hardware in the IR output. The IEEE standard requires that IR output bits [1:0] be 0,1 respectively. The remaining bits can
reflect status information.

At power-up or during normal operation, TRST can be asserted to initialize the
test controller. This immediately places the TAP in the Testlogic-reset state. The
TAP can also be forced into the Testlogic-reset state by driving TMS high for five
TCK cycles. Five is the maximum number of TCK cycles required to transition the
TAP to the Testlogic-reset state from any of the other states when TMS is held high.
In the Testlogic-reset state, the TAP issues an internal reset signal that places all test
logic in a condition that does not impede normal operation. The TAPS also locks
the IDCODE instruction into the instruction register and selects the device ID register as the default data register at reset.

From the Testlogic-reset state, the TAP moves to the Run test/idle state when
TMS is pulled low. As long as TMS is held low, the TAP stays in the idle state.
From this state, driving TMS high moves the TAP to the data register scan cycle.
The TAP cannot remain in the select DR scan state for more than one TCK cycle.
Driving TMS low for one TCK cycle causes the TAP to begin the data register scan
process, moving to the capture DR state. Keeping TMS high for one more TCK
cycle moves the TAP to the beginning of the instruction register scan cycle (select
IR scan state).

After reset, one can read the device ID register (default). To perform any other
action, one must move the TAP to the instruction register scan cycle to select an
appropriate data register. For either type of scan cycle (data register or instruction
register), the first action in the scan cycle is a capture operation. The capture-DR
state enables the data register indicated by the current instruction register contents.
The capture IR state enables access to the instruction register.

From the capture state, the TAP transitions to either the shift or the exit1 state. The shift state allows test data or a new instruction to be shifted in or status information to be shifted out for inspection. Following the shift state, the TAP either returns to the Run test/idle state, via the exit1 and update states, or enters the pause state, via exit1. The pause state allows data shifting through either the selected data register or instruction register to be temporarily suspended while a required operation is performed. From the pause state, shifting can resume by re-entering the shift state via the exit2 state, or it can be terminated by entering the Run test/idle state via the exit2 and update states.

3.3   JTAG Registers

All registers are accessed serially through the TAP and, when selected, connect between the TDI and TDO pins. The TAP controller, which is a state machine, controls access to the registers. The state is changed by the TMS signal in conjunction with the TCK.

The following registers are found in most JTAG systems:

- The bypass register provides a single-bit scan path between TDI and TDO. It enhances test efficiency when a device other than the core-based device becomes the device under test. When the bypass register is selected by the current instruction, the shift register stage is set to a logic 0 on the rising edge of TCK in the capture DR controller state. Therefore, the first bit shifted out after the bypass register is selected is always 0.
- The ID register is a 32-bit register that stores values that identify the device manufacturer, part number, and version of a device and is selected by the IDCODE instruction. It can be used to distinguish specific IEEE 1149.1–compliant parts in a daisy-chained system. The least significant bit (bit 0) is always set to logic 1, as required by the standard; this bit is an identity packing bit that indicates valid data.
- The boundary-scan register defines test operation in the device and contains bits for all signal, clock, and control pins. All bidirectional pins have a single register bit and an associated control bit in the BSR. In the update DR state, the register contains valid stimuli data. In the capture DR state, the boundary-scan register samples data. Data clocked into the device in the shift DR state can drive output pins in the subsequent update DR state. At the same time, the clocking action shifts out sampled pin data from the previous capture DR state.
- The instruction register is a required register specified in IEEE Standard 1149.1 that must be at least 1 bit long. Different processor families implement different-length IRs that decode the unique instructions supported for the device. For instruction operation codes that are not defined, the standard requires that the decoder select the bypass register by default. The IR consists of a shift register with parallel outputs. Data transfers from the shift register to the parallel outputs during the update IR TAP controller state. During a shift IR loading sequence,

data can be clocked through the instruction register out of TDO to allow instructions to be passed to any subsequent devices in the JTAG daisy chain. During the capture IR state, the parallel inputs to the instruction shift register are loaded with 01 in the least significant bits as required by IEEE Standard 1149.1. The two most significant bits are loaded with the values of the core status bits [1–0] from the debug controller.

After reset, the instruction register is loaded with the IDCODE instruction, and the ID register is the selected data register. One can perform a data scan to read the device information. For other operations, the TAP programming sequence must begin with a scan into the instruction register to specify the appropriate data register. After an instruction register scan, subsequent scans are through the specified data register and may involve several scans of data into or through it.

Debug registers in most cases are data registers that are accessed through a command register that is configured during an IR pass by loading a debug-related instruction.

3.4 JTAG Instructions

With the exception of BYPASS and IDCODE, which are defined and discussed in later chapters, none of the mandatory or optional instructions referenced in the 1149.1 specification are used for debug purposes and are of no direct interest to this discussion. It should be noted, however, that BYPASS, EXTEST, and SAMPLE/ PRELOAD instructions are mandatory and, as such, should be included in a TAP design if it is intended to be 1149.1-compliant.

BYPASS (11...11): This instruction is required by IEEE Standard 1149.1 and is defined to be all 1s. BYPASS allows the device to remain in its functional mode and connects the bypass register between TDI and TDO. It allows serial data to pass through from TDI to TDO without affecting the operations.

IDCODE: This optional instruction is specified in IEEE Standard 1149.1. IDCODE allows the device to remain in its functional mode and connects the ID register between TDI and TDO. It allows the user to read the manufacturer, part number, and version of a component from the TAP. This is the default value loaded into the IR at reset.

Debug-related instructions are discussed in the next chapters, both generically and for various commercial and widely used debug instruments.

JTAG has therefore evolved into a robust leading mechanism for debug control even though, due to its serial architecture, JTAG it is limited in the level of debug visibility it can support. JTAG works well for debug of a single processor in isolation because debug may be based on observation of a limited number of internal registers and the processor may be halted to probe and export additional information. The increased signal complexity of emerging SoC devices, with multiple processing operations distributed over many resources and communication with other supporting IP and internalized buses, in some cases require data access that outstrips JTAG bandwidth. Debug interfaces have been developed to provide complementary instrumentation interfaces that allow higher levels of data throughput than JTAG. These are discussed in later chapters.

3.5 Boundary-Scan Description Language

Boundary-scan description language (BSDL) describes how IEEE 1149.1 is implemented in a device and how the device operates. A BSDL description for a device is based on VHDL model descriptions and consists of the following elements:

- An entity description.
- A generic parameter.
- A logical port description.
- Use statements.
- Pin mapping(s).
- A scan port identification.
- An instruction register description.
- A register access description.
- An ID code description.
- A boundary register description.

BSDL does not describe how instruments are controlled by the JTAG TAP operations. A BSDL description of a device consists of at least an entity description, a generic parameter, a logical port description, pin mapping, a scan port identification, an instruction register description, an ID code description, and a boundary register description.

For those not familiar with VHDL syntax, descriptions are provided for each stage of the file.

BSDL code is in Courier font, *comments* are shown in italics.

– BDSL HEADER - *This is free format* and typically not parsed
– Per VHDL Syntax, lines starting with – are comments
– Showing Boundary Scan Description Language (BSDL) for
– Device DSP_NS in a 24-pin package
– Modification History
– Date Author Version
– Created –/–/–' NSTOLLON 1.1
– Modified –/–/–,

The entity description gives the name of the device. It begins with an entity statement and terminates with an end statement. For example, this entity description for device called DSP_NS
    entity DSP_NS
is a generic parameter that can come from outside the entity, In BSDL, the only generic is a string with a name PHYSICAL_PIN_MAP. If "undefined" it can be

defined by another file, or it can be defaulted to a fixed value such as the package type, in this case, a 24-pin package.

```
generic (PHYSICAL_PIN_MAP: string:= "PQ24");
```

The logical port description gives logical names to the I/O pins and specifies whether the signals are input, output, bidirectional, or linkage (for power supply).

```
port (
P1:in bit;
P2:inout bit;
VSS:linkage bit;
XIN:linkage bit;
XOUT:linkage bit;
VCC_BUS:linkage bit;
P3:inout bit;
P4:inout bit;
P5:inout bit;
P6:inout bit;
P7:inout bit;
VREF:linkage bit;
P8:inout bit;
EXC_VDD:linkage bit;
P9:inout bit;
RESET:in bit;
MOD0:in bit;
MOD1:in bit;
VDDE:linkage bit;
TMS:in bit;
TCK:in bit;
TRST:in bit;
TDO:out bit;
TDI:in bit
);
```

STD 1149_1_1994 refers to a package of predefined functions and components that are associated with various attribute statements referenced in the BSDL model. A STD_1149_1 use statement is mandatory in BSDL. The ".all" suffix means to use all components and functions of the package.

```
use STD_1149_1_1994.all;
attribute COMPONENT_CONFORMANCE of DSP_NS: entity is
"STD_1149_1_1994";
```

For example, PIN_MAP is a predefined function in the STD 1149_1_1994 package that maps logical signals to the physical pins of the 24-pin package.

```
attribute PIN_MAP of DSP_NS: entity is PHYSICAL_PIN_MAP;
constant PQ24: PIN_MAP_STRING:=
"P1:1," &
"P2:2," &
"VSS:3," &
"XIN:4," &
```

```
"XOUT:5," &
"VCC_BUS:6," &
"P3:7," &
"P4:8," &
"P5:9," &
"P6:10," &
"P7:11," &
"VREF:12," &
"P8:13," &
"EXC_VDD:14," &
"P9:15," &
"RESET:16," &
"MOD0:17," &
"MOD1:18," &
"VDDE:19," &
"TMS:20," &
"TCK:21," &
"TRST:22," &
"TDO:23," &
"TDI:24,";
```

Attributes for scan port identification statements define the TAP, device clock, and reset operations. The definition of the port of a device contains four mandatory pins (TDI, TDO, TCK, TMS) and one optional TRST pin.

```
attribute TAP_SCAN_IN of TDI: signal is true;
attribute TAP_SCAN_MODE of TMS: signal is true;
attribute TAP_SCAN_OUT of TDO: signal is true;
attribute TAP_SCAN_CLOCK of TCK: signal is (5.0e6, BOTH);
attribute TAP_SCAN_RESET of TRST: signal is true;
```

The instruction register description identifies device-dependent characteristics of the instruction register, including the device-specific instructions supported by a given device. These include debug-related instructions. In this example, the instruction register length attribute defines the instruction register length as 6 bits and gives the instruction opcode definitions. It also specifies that for the capture IR state parallel inputs to the instruction shift register are loaded with value 110001.

```
attribute INSTRUCTION_LENGTH of DSP_NS: entity is 6;
attribute  INSTRUCTION_CAPTURE  of  DSP_NS:  entity  is
"110001";
attribute INSTRUCTION_OPCODE of DSP_NS: entity is
"BYPASS (111111)," & - defined as required instruction
by JTAG
"SAMPLE (000001)," & - defined as required instruction
by JTAG
"EXTEST (000000)," & - defined as required instruction
by JTAG
"IDCODE (000010)," & - defined as optional instruction
by JTAG
```

```
"USERCODE (000011)," & - other are example user-defined
instructions
"DBG_SYSTEM    (001000)," &  -  for  debug  control
operations
"DBG_CONTROL (001001)," &
"DBG_SETUP (001010)," &
"MON_CONTROL  (001111)," &  -  for  debug  monitoring
operations
"MON_CODE (010000)," &
"MON_DATA (010001)," &
"MON_PARAM (010010)," &
"MON_ACCESS (010011),";
```

The attribute INSTRUCTION_PRIVATE register description identifies user-defined instructions. It does not include features to describe their functions.

```
attribute INSTRUCTION_PRIVATE of DSP_NS: entity is
"DBG_SYSTEM," &
"DBG_CONTROL," &
"DBG_SETUP," &
"MON_CODE," &
"MON_DATA," &
"MON_PARAM," &
"MON_ACCESS";
```

The ID code register description identifies the values captured in the device identification register when the IDCODE instruction is executed.

```
attribute IDCODE_REGISTER of DSP_NS: entity is
"0000" & - version
"0011001000100110" & - part number
"01000100011" & - manufacturer's identity
"1"; - required by 1149.1
```

Additional attributes can be defined for other registers.

```
attribute DBG_SYSTEM_REG of DSP_NS: entity is
"0000000000100000" & - reserved
"0000" & - trigger modes
"0001" & - ROM monitor
"0010" & - ISA debug
"0100"; - debug state
```

A register access description defines the name of a register placed between the TDI and TDO for each instruction.

```
attribute REGISTER_ACCESS of DSP_NS: entity is
"Bypass (BYPASS)," &
"Boundary (SAMPLE, EXTEST)," &
"DEVICE_ID (IDCODE,USERCODE)," &
"DBG_SYSTEM_REG (DBG_SYSTEM)," &
"DBG_CONTROL_REG (DBG_CONTROL)," &
"DBG_SETUP_REG (DBG_SETUP)," &
"MON_CODE_REG (MON_CODE)," &
```

```
"MON_DATA_REG (MON_DATA)," &
"MON_PARAM_REG (MON_PARAM)," &
"MON_ACCESS_REG (MON_ACCESS);
```
The boundary register description lists the boundary-scan cells and gives information regarding the cell type and associated control.
```
attribute BOUNDARY_LENGTH of DSP_NS: entity is 27;
attribute BOUNDARY_REGISTER of DSP_NS: entity is
"26 (BC_4, P100, observe_only, X)," &
"25 (BC_6, P100, bidir, X, 13, 1, Z)," &
"24 (BC_1, *, control, 0)," &
"23 (BC_4, RESET, observe_only, X)," &
"22 (BC_4, MOD0, observe_only, X)," &
"21 (BC_4, MOD1, observe_only, X)," &
"20 (BC_4, P101, observe_only, X)," &
"19 (BC_6, P100, bidir, X, 13, 1, Z)," &
"18 (BC_1, *, control, 0)," &
"17 (BC_4, P102, observe_only, X)," &
"16 (BC_1, P102, output3, X, 0, 0, Z)," &
"15 (BC_1, *, control, 0)" &
"14 (BC_4, P103, observe_only, X)," &
"13 (BC_1, P103, output3, X, 291, 0, Z)," &
"12 (BC_1, *, control, 0)," &
"11 (BC_4, P104, observe_only, X)," &
"10 (BC_1, P104, output3, X, 288, 0, Z)," &
"9 (BC_1, *, control, 0)," &
"8 (BC_4, P105, observe_only, X)," &
"7 (BC_1, P105, output3, X, 285, 0, Z)," &
"6 (BC_1, *, control, 0)," &
"5 (BC_4, P106, observe_only, X)," &
"4 (BC_1, P106, output3, X, 282, 0, Z)," &
"3 (BC_1, *, control, 0)," &
"2 (BC_4, P107, observe_only, X)," &
"1 (BC_1, P107, output3, X, 279, 0, Z)," &
"0 (BC_1, *, control, 0),";
end DSP_NS
```

BSDL is primarily a means to describe 1149.1 operations, is not structured to be very useful for on-chip instruments by itself, because too much instrument functionality is outside of the 1149.1 standard. BSDL is discussed in Chap. 9 (IJTAG) in the context of a litmus test for JTAG components. If it is 1149.1-compliant, it should be describable in BSDL. Conversely, if it cannot be described in BSDL, it is not 1149.1-compliant. There are activities to define successor languages to BSDL to encompass features for more in-depth description of instrumentation operations, configurations, and on-chip debug-related functionality.

3.6   The Road to JTAG: Historical Debug Approaches

The majority of on-chip debug interfaces in use today are based on JTAG. It is useful to discuss some of the previous approaches, because many, like emulation, have continued to evolve into related but somewhat separate disciplines, and others, like BDM, which was in wide use a decade ago, continue to see some use today. Given the central importance and relative complexity of processor cores in embedded systems, a majority of the focus has been on processor debug technologies rather than on debug-related aspects of an embedded system.

Tools that to a large extent only address the specifics of the processor are obviously limited in more system-related applications. As the processors become more deeply embedded, traditional development tools for system debug applications cannot provide nonintrusive visibility into the highly integrated embedded processor. Applied to processor in-circuit emulators and their derivatives, the system must be placed in special debug modes or halted before it can probe processor registers or read/write to the embedded memory. In many cases, this interruption of the steady-state performance of the system introduces (time) intrusive elements into the system operation that can complicate or invalidate the data or operations being observed. This problem grows proportionally to the ever-increasing frequency and complexity of high-performance embedded processors.

Printf-based debug: Historically, the most commonly used processor debug tool is some variation of the printf command, which allows the processor, during its normal course of operation, to transfer status information to either memory or an external interface for later interpretation of the program operation and hopefully some signposts of where and when errors are occurring. It is likely that even today, variants of printf commands embedded in code running both in simulations and on hardware targets are the most widely used means of system debug. The disadvantage of using printf commands in embedded code is the different ways that embedded compilers support printf commands and the fact that by adding these statements to one's code, one changes the program flow that one is trying to debug. There are software-related books that address the use and variants of the printf command for use in software debug, so we may safely leave this topic and the interested user can find many alternate sources of information.

Debug monitors: A related approach is the use of a debug monitor (or a ROM monitor as it was often included in the boot ROM of a device), which is code that is included in the processor to help with debugging. It usually communicates via a serial interface to a host computer or some form of terminal. A basic monitor allows for the download of code, the reading and writing of memory and registers, and, perhaps most importantly, setting breakpoints, single stepping, and real-time execution. More complex monitors allow source code profiling and complex breakpoints. A variant of the debug monitor uses a ROM emulator as a plug-in replacement for the on-chip or on-board ROM containing the debug monitor code. The ROM emulator device would typically be connected to a host computer to allow debug code download (as opposed to having dedicated debug support in ROM) that contains a ROM monitor and communicates with the monitor via the emulator interface, as opposed to having dedicated debug support.

In a systems analysis context, however, monitors present debug limitations. Debug monitors are intrusive into operational flow of the processor, change the state of the processor, and control the program's execution by changing the memory map of the processor to accommodate inclusion of the monitor code and forcing the processor to always be on. Debugging of interrupts and real-time operations are thus typically not feasible. Breakpoints are typically limited to those implemented in software (inserting an opcode for a "trap"), further changing the system being debugged. Single stepping is often done by inserting breakpoints in appropriate places.

In-circuit emulation: The in-circuit emulator (ICE) has had a long and generally successful history for stand-alone processor debug devices. In the 1990s, it was virtually required that every new device have an ICE system to be used for debug and systems integration. ICE typically used a special version of the processor called a "bond-out" chip with extra debug support pins, bringing typically internal signals to the chip periphery in order for code execution to be monitored and traced, the processor controlled using hardware triggers and breakpoints, and external memory to be mapped into the user space. Due to architectural differences in processors, diverse ICE tools have been developed and continue to evolve to suit the needs of different processors.

The ICE concept continues to be used, with an entire electronics industry subset dedicated to emulators that replace the bond-out chip with a programmable implementation that allows the processor functions along with other system logic to be implemented and executed. However, these emulators run on a different principal, essentially replacing entire systems with their FPGA equivalents.

In ICE operations, the processor operational interfaces are typically modified, via either software or hardware, to allow extended host control of the processor run environment. When the processors are in ICE mode, they may be in a non-standard operating state of the processor with different opcodes and interface features. The resulting operation of a processor in emulation mode makes the processor appear to be in a dormant state, with impact on its ability to access and debug other parts of the embedded system. In addition, to be minimally intrusive, many processor emulation schemes are limited to monitoring the processor bus. Many signals and internal registers may be inaccessible during ICE mode operations, although this may not be the case for more general emulation. Other limitations include the inability to debug at full speed and concerns for subtle differences in operation between an emulated version of a processor and the actual processor.

Most in-circuit emulators contain real-time trace circuitry, which allows them to capture the activity on the processor's bus and, with on-chip support, the processor's internal states. This data is generally logged to a trace buffer for later analysis. Such data is particularly helpful when trying to debug problems involving behavior that can only be captured when the processor is running at full speed.

The ICE's most powerful features include complex breakpoints (even in ROM), real-time traces of processor activity, and no use of target resources. But this extra functionality requires separate packaging and, in some cases, a separate die. The speed, complexity, and integration levels of modern processors limit the availability and feasibility of bond-out versions, making emulators difficult and expensive to design. As a result, some debugging features unique to ICEs are not available for modern processors.

3.6.1 Background Debug Mode

Background debug mode (BDM) is a bit-serial synchronous communication developed by Motorola. The debug signal interface consists of a serial data in, serial data out, serial clock/breakpoint, and a freeze status signal. At its most basic implementation, BDM allows externally controlled read or write of a range of registers and on-chip memory. There are several BDM variants that also allow a BDM interface to set a break or interrupt to debug mode under varying conditions, to halt execution of normal machine code fetched from the memory, and to start to process commands received from the serial debug interface.

BDM commands are similar to those in ROM monitors. Single stepping is accomplished by hardware control of the BDM port or by placing a software breakpoint instruction in the code stream. Although BDM is still in limited use, it is primarily interesting as the first example of a debug port whereby commands can be used to view and modify registers and to access on-chip and external memory locations.

The basic BDM command set is generally the same across processor families, but differences exist due to the inherent architectural differences. These differences are handled by the particular debugger that drives the BDM.

BDM commands are 17 bits long (actually 16 command bits and 1 status/control bit). Commands are shifted serially along the serial-data-in (DSI) signal from the debugger to the processor; each may be followed by one or more extension words. Responses are shifted serially out of the processor on the corresponding serial-data-out (DSO) signal. These data transmissions are synchronized to a serial-clock (DSCLK) signal, which is driven by the remote debugger. We can see how JTAG was identified as a logical successor to BDM, because it shares some of the same architectural concepts.

Table 3.2 shows a core BDM command set. Commands are similar to those of a typical debug monitor. An external debugger host is enabled for capabilities like reading and writing registers and individual memory locations. BDM commands invoked while the processor is running involving memory will "steal" bus cycles from the processor, much as a DMA (Direct Memory Access) controller would.

The debugger performs memory and register read/write and processor halt/restart operations, without the processor involvement or impact to instruction flow when these activities are occurring. Execution of a background mode (BGND) instruction

Table 3.2 A basic BDM command set

Command	Mnemonic	Description
Read register	RAREG/RDREG	Read the selected address or data register and return the result
Write register	WAREG/WDREG	Write the specified value to the selected address or data register
Read memory	READ	Read from the specified memory location
Write memory	WRITE	Write to the specified memory location
Dump memory	DUMP	Read from a block of memory
Fill memory	FILL	Write to a block of memory
Resume execution	GO	Resume instruction execution at the current value of the PC (after pipeline flush)

or assertion of a breakpoint signal from the debugger will cause the processor to halt and the on-chip debug hardware to perform operations until a command to resume normal execution (GO) command is received.

BDM, like other on-chip debuggers, provides basic capabilities similar to a debug monitor, but debugging does not need to use target memory. It also offers some of the features of an ICE to view registers and memory without halting the processor.

On-chip debug instruments allow users to see address and data values just as the processor sees them, that is, unfiltered by pre-fetch or cache operations. In a production system, it is only possible to capture them with an on-chip debugger. For example, Freescale ColdFire's BDM connection contains eight additional output signals, which can output nibble-formatted information on the processor's state. By logging data on the host side, the real-time execution history of the processor can be reconstructed from this information.

Chapter 4
Processor System Debug

Debug features for embedded processors have been recognized from the earliest days of embedded processing as an important requirement for processor verification. Because detailed simulation of processor operations for many applications has historically not been feasible due to the large number of cycles required for many applications, processor analysis via emulation and trace of processor operations has been required for verification and hardware/software integration. Most licensable embedded processors include some instrumentation features to support debug. Although the specifics vary with each processor type, debug for processor cores typically provides similar debug features:

1. Processor-specific run control (start, stop, software and hardware breakpoints, single-step run control).
2. Monitoring of hardware and software breakpoints for triggering.
3. Real-time trace that can include execution (instruction) and/or data trace. Trace operations can be triggered from conditions such as instruction execution, memor, or IO operations, address range, or opcode value.

Among the most valuable processor debug features for analyzing operational performance is execution trace. Trace in general is a complex debug technology because it requires either a large buffer or high bandwidth to export trace information. Each new generation of processors brings new performance capabilities that make debugging more difficult. To address these new barriers, processor manufactures have been adding parallel debug capabilities to devices, enabling a new class of debugging techniques that promises to help developers get home on time Tables 4.1 and 4.2.

Integrating debug instruments on processors allows JTAG-driven emulation and pseudo-real-time debug through access to system registers. These debug instruments enhance the visibility a JTAG port can provide into processor operation. Instead of using the processor core to execute functions, debug peripherals execute in parallel to the processor with complete access to system registers, memory, and executive control, resulting in nonintrusive visibility, increased performance, lower latency, and greater complexity of functions.

N. Stollon, *On-Chip Instrumentation: Design and Debug for Systems on Chip*,
DOI 10.1007/978-1-4419-7563-8_4, © Springer Science+Business Media, LLC 2011

Table 4.1 Instrumentation chip IO interfaces

Name	Type	Description
		Initialization and clocking
CLK	Input	Clock source
RESET	Input	Hardware reset input. Clears internal OAI resources
		JTAG
TCK	Input	Test Clock input. Asynchronous to but lower frequency than CLK
TMS	Input	Test mode select. Comes directly from input pad
TDI	Input	Test data input. Comes directly from input pad
nTRST	Input	Test reset. Active high
TDO	Output	Test data output. Goes directly to output pad
		External debug mode control
BreakIn	Input	Input signal from cross-trigger bus (coming from other internal / external logic)
BreakOut	Output	Output signal to cross trigger. Driven high on processor breakpoint (going to other internal/external logic)
		External trigger
Trig_Out	Output	Trigger out connects to cross-trigger logic or output pin allows cross triggering with other logic

Table 4.2 Instrumentation processor interfaces

Name	Type	Description
		Processor debug mode control
DebugAck	Input	Debug acknowledge. Connected to processor debug_ack output
DebugReq	Output	Debug start/stop request. Connected to processor debug request input
DebugStep	Output	Debug step. Connected to processor debug_step input
Debug_Prog	Output	Debug operation to inhibit PC when debug instrument-driven read/write operation is active
		Processor trace/trigger
Fetch	Input	Program fetch
Flush	Input	Program branch indicator. Connected to processor flush output
Memaddr	Input	Program memory bus. Connected to processor memaddr bus

On-chip processor instrumentation may be added to processor cores, providing run control, memory and register visibility, complex breakpoints, and trace history features. Typically, analysis in many processors have common features.

- Control via 4-pin IEEE-1149.1 (JTAG) port.
- Start/stop run control through DebugReq and DebugAck handshake signals to core.
- Support for an unlimited number of software breakpoints using a software breakpoint opcode.
- Single-step operation by assembly instruction.
- Access to registers and memory (code, external data, SFR, and internal data spaces) while user code is running with minimal impact to real-time performance.
- A fixed or scalable number of hardware breakpoints consisting of an address/ data value under different modes of operation, including:

- Code memory execution.
- Code memory read or write.
- External data memory read or write.
- SFR read or write.
- Internal data memory read or write.

Some processors allow combinations of two hardware breakpoints to form an
address range (lower and upper bounds) and masked data value. Hardware breakpoints may be configured to enter or leave emulation or debug modes, start or stop
a trace operation, or assert signals to change bus or register values or trigger
outputs.

More advanced processor debug allows:

- Cross-triggering for multiprocessor synchronization.
- Trace history of the most recent branch points for software reconstruction of
  execution flow. Branches record both branch-from and branch-to addresses.
  Trace start/stop triggers allocate the trace frame
- Support for multiple-memory-bank systems in breakpoint decisions and trace.
- Support for code memory, external data memory, and SFR trace.

A key element is that they perform these functions in a way that does not impact
processor performance significantly.

To show some examples of debug instrumentation operation, we can construct
a simple generic processor with an on-chip debug instrument having a trace
capability. Off-chip interfaces for the instrumentation are JTAG, trigger, and
break signals. On-chip interfaces are to the processor core and RAM as shown
in Fig. 4.1.

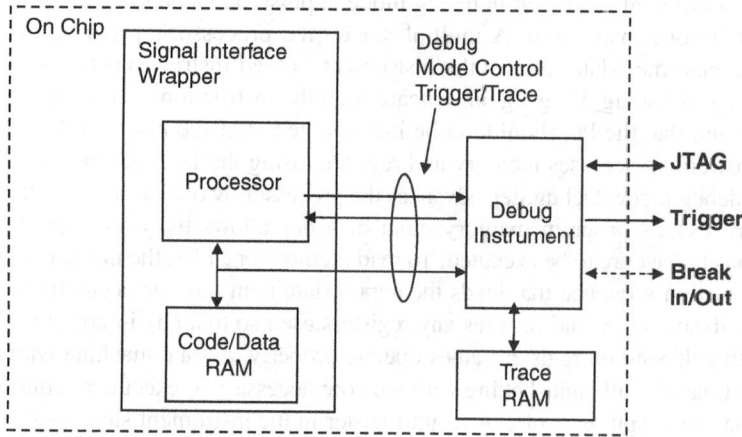

Fig. 4.1 Processor on-chip analysis instrumentation

4.1   A Processor Debug Instrument Implementation

During initialization, the instrument samples the system RESET signal to initialize to a known state. The instrument samples the state of TCK at the trailing edge of RESET. If TCK is sampled high, the instrument concludes that the debugger is not installed and does not affect normal processor operation. If TCK is low, the instrument holds the processor in the reset state until the external debugger can assert a DebugReq via the JTAG port.

Start and stop execution are both handled by a DebugReq signal. To start execution, the instrument asserts DebugReq. To stop execution, the instrument asserts DebugReq. When the current instruction is completed, the core enters debug mode and returns an acknowledgment by asserting DebugAck.

Single-step: From debug mode, the instrument executes one user instruction by pulsing the Debug_Step signal active for one clock. The processor responds by fetching and executing one instruction, then returning to debug mode. DebugAck is negated during the step.

Software breakpoints: Processors use some variant of a TRAP signal to trigger a software breakpoint. After execution of the processor TRAP instruction, the core switches to debug mode and asserts DebugAck. Through the JTAG port, the debugger system periodically polls DebugAck and begins breakpoint processing when it becomes asserted.

Fielding user interrupts in debug mode: The processor signals to the instrument that a user interrupt is pending. The instrument completes any operation in progress and then negates DebugReq, waits for DebugAck to indicate that debug mode has ended, and then reasserts DebugReq. When the processor completes the interrupt service routine, it performs a return from interrupt routine and returns to debug mode at the original PC. The sequence is identical to that for single-step except that the process is initiated from a user signal rather than from the external hardware.

Memory/SFR access: The instrument accesses memory and registers using the DebugStep mechanism. When in debug mode, a pulse on DebugStep advances the processor by one instruction. A multiplexer on the processor's program memory input data bus (memdata) allows the instrument to feed instructions to the core to be executed. A Debug_Prog signal indicates that the instruction is coming from the debugger and that the PC should not be incremented after the instruction.

The instrument accesses memory and registers using the DebugStep mechanism. When in debug mode, DebugStep advances the processor by one cycle. A multiplexer on the processor's program memory input data bus allows the instrument to feed instructions to the core to be executed. To read memory or SFRs, the instrument feeds in an instruction sequence that loads the appropriate item into the accumulator. The sequence always saves and restores any registers used so that any interrupt handlers invoked that depend on register values operate properly. A state machine within the instrument handles all handshaking with the core necessary to execute a sequence of instructions and capture results. A result register in the instrument stores state information. The result is available to read via JTAG once the sequence is completed.

Processor trace: A scalable trace buffer memory stores branches executed by the core. At every change of flow, the most recent PC from the old code sequence and the first PC from the new sequence are stored together as a trace record. Change of flow events include branches, calls, returns, interrupts, and reset. Two signals provided by the core assist in collecting this information: fetch is active when the core fetches program code, indicating that the current PC is present on the PROGA bus, and flush indicates that the program fetch in progress is the first from the new code sequence.

When trace begins, the trace memory address and a trace wrap flag are initialized to zero. As each record is stored, the address increments, wrapping back to zero when the memory is filled. A status bit is set when the trace address wraps.

The instrument maintains a shadow of the core's PC by writing the memaddr bus into a holding register at every assertion of fetch. Upon sampling flush active, the instrument writes the PC holding register and the address from the current bus cycle in successive clocks, incrementing the trace address.

When tracing stops due to a trigger or breakpoint, the trace memory is frozen and the trace memory pointer indicates the first unused memory location. If the trace wrap flag is set, the first frame collected is the one currently addressed and the last collected is the one just preceding it. If the trace wrap flag is not set, the oldest frame is frame zero and the youngest is the one just preceding the trace address. After a breakpoint, the trace memory contents are read out through JTAG for display.

Processor triggers: The instrument provides a set of hardware breakpoint or trigger registers that monitor bus activity and perform various actions when specified bus events occur.

Each trigger register is accessible through JTAG. There are three fields in a trigger register: address, data, and status, each with its own enable bit and mask field. This allows triggering based on address only, data only, and address in combination with data with processor status bits optionally participating in trigger decisions.

A trigger mode input allows selection of different buses' code read/write, data read/write, instruction data read/write, or SFR read/write. These bits direct the appropriate bus signals to the trigger comparator fields:

- When all three fields of the trigger are satisfied, an action occurs. The trigger register selects which action(s) to perform for each trigger. The actions possible are asserting DebugReq, (i.e. break emulation), asserting Trig_Out, and starting or stopping trace collection.
- Trig_Out is an optional output signal from the instrument to either a pin or another on-chip logic. The Trig_Out then connects to the external debugger and is available for external monitoring. Similarly the Trig_In signal is input from external logic over a pin or from another on-chip logic. Systems can have more than one Trig_In and Trig_Out signal that are controlled independently.
- Code execution breakpoints are different from other modes in that they do not perform any action directly. Instead, they override user code feeding the processor with the TRAP instruction. If the instruction is executed, then a breakpoint occurs. This allows breakpointing of code read from read-only memory.

- In systems where two or more triggers are implemented, pairs of triggers may be combined to form a "super-trigger" by setting the combination bit in each of the trigger registers of the pair. Trigger registers are combined in fixed pairs. Different combination modes may be defined, either statistically or dynamically; for example, in combinatorial event mode, a trigger pair is satisfied when, as for the following example; the address is in a range defined by Trigger0 and Trigger1, data meets some algebraic condition, or status is at some defined value.

Trigger0.Address <= TrigAddr <= Trigger1.Address
AND (Trigger0.Data XOR Trigger0.Mask) AND Trigger1.Data == 0
OR Status == ABCD;

The address must be between a lower and upper bound denoted by the address fields in Trigger0 and Trigger1 respectively, and the data, masked by the data field in Trigger1, matches the data field in Trigger0, or the status matches the defined value.

One can also create inverted breakpoints. Say one have a variable being modified in the code. By defining an inverted range – any code outside the function rather than inside it – one will narrow the number of modifications one has to personally evaluate, increasing overall efficiency.

A cross-trigger interface is intended to interconnect two or more processors so that when any one processor hits a breakpoint or trigger condition, all others are requested to break or take other action (as defined by the cross-trigger logic) within a few clocks.

Debug instruments can also perform writes and reads without halting the processor (also called real-time data exchange (RTDX)), allowing polling of registers and memory address ranges as the application code executes. This feature can be useful in generating real-time errors or tracking program execution. RTDX can be a real-time feature; the read/write can be made based on a precise trigger. What is important is that real-time events are not affected by the read/write. This is critical for applications servicing real-time events. For advanced program execution tracking, one can watch the program counter or instrument the code to adjust debug variables that describe the current status of the application. For example, when one set a breakpoint, the instruction opcode is replaced with a breakpoint opcode that halts or otherwise takes action such as initiating an interrupt on the processor.

In these types of cases, the option to continue to execute interrupts even when the application is halted can be quite useful. This is achieved with an embedded emulation peripheral that masks time-critical interrupts. Being able to mask interrupts is important because there may be certain application-based timer interrupts one don't want executing.

To do this, one need to make the task manager aware of the time-critical interrupt mask. When determining whether to begin the next task, check if the system is in jeopardy. If it is, execute the task. If not, then one can halt the task queue. If the task queue is empty, the task manager must queue a task that removes the system from jeopardy. When the processor resumes execution, the task manager returns the system back to the condition it was in when the queue was empty.

Time-critical interrupt masking can also simplify hardware debugging. Consider a one-second action. On a 100-MHz processor you'll have to hunt down the small amount of real-time code interspersed among approximately 100 million lines of application code. Using time-critical interrupt masking, you could freeze the task manager until the tasks you want to debug are queued. If you mask for time-critical interrupts and release the task manager, the processor will be halted for the application but will still run the real-time code. Thus, all you'll have in the trace buffer is the real-time code that you want to debug. Of course, if the bug is caused by an unintended interaction between the application and interrupt, this technique will not reveal the problem. However, you will know it is not solely the interrupt at fault but rather an unintended interaction.

4.2 Processor Trace Compression

To perform an instruction trace, one must first set a trigger point. Unlike a simple trigger point of a particular instruction address, a complex trigger point may involve counters, logical operators, bit masking, and event sequencing; there are two ways to generate a trigger point:

1. Use the hardware debug resources contained in the processor core.
2. Use an external trigger source to feed into the processor core.

Using hardware debug resources: Most RISC processors have registers and debug facilities that allow users to set breakpoints at different instruction pointers, at the address of one or more data at addresses, when a branch or exception is taken, etc. For real-time instruction tracing, these internal processor core resources are used to determine the trace point, instead of a breakpoint, in order that the processor flow is not halted.

This method has the advantage of being a precise trigger mechanism, which means that the exact point when the trigger executes is known. In this method, all code before the trigger point is guaranteed to have been executed, and all code after the trigger point has not yet been executed.

Using only available processor resources for trigger points may sound limiting, but it is not a problem. Because processor cores incorporate more and more debug resources to help the embedded developer, it is not a problem. Most RISC processors contain breakpoint/trigger points for multiple instruction address and data address values, as well as counting and sequencing mechanisms for when branches or exceptions occur.

Using an external trigger source: If the internal hardware resources are not sufficient, an external trigger can be fed into the processor to be used as the trigger point. This method is an imprecise triggering mechanism, because the event has already occurred before being fed into the processor core. It is therefore likely that processor execution has continued past the trigger point (this is also known as "skid," because the pointer skids past the desired trigger point). Fortunately, the amount of skid is usually minimal and does not hinder the usefulness of the instruction trace.

Compressing the trace data: The trick is to get the necessary instruction address information onto seven data pins. Fortunately, the locality of reference associated with Von Neumann architecture machines assists in this process. The following explanation refers to the number of finite "states" needed to determine the code flow, plus some dedicated pins for special address broadcasting.

Linear code execution: Consider the normal, linear (sequential) execution flow of a scalar 32-bit RISC architecture. Linear code flow means that after an instruction is executed at an address pointer, the next instruction to execute is located at pointer + 4 (assuming a 32-bit instruction width). To determine instruction address flow, we must broadcast two states during every processor core clock period – one state to say that an instruction has executed on a given clock cycle, and another to say that no instruction executed on the given cycle. To illustrate, let us look at the following example. Consider the following trace, in which State 0 means the instruction did not finish executing on the given cycle, and State 1 indicates the instruction finished executing on that cycle:

Some reasons an instruction may not execute on a cycle are: multicycle instructions (such as multiplies and divides), memory accesses, and pipeline stalls. As one can see from Table 4.3, only one data pin is needed to save the two states. Assume, for the purposes of an example, that we already know that the beginning pointer is at 0×10 when the trace is started.

Because the code flow is known to be linear, the order of the addresses will always be increasing where the next pointer = current pointer + 4. The only remaining calculation is to determine how long each instruction took to execute. This is calculated by adding the number of nonexecuted cycles plus the cycle the instruction did execute. For example, at the beginning address, 0×10, the instruction took two cycles to execute, because it did not execute on Cycle 1 but did on Cycle 2. A postprocessing tool would determine that the instruction flow was the following: When the trace was gathered, memory or a static code listing can be read to determine the instructions at addresses $0 \times 10, 0 \times 14, 0 \times 18$.

In this six-cycle example, only one bit of information per cycle must be saved, for a total of six bits rather than the 32 bits per cycle needed if we were storing the pointer itself on every cycle. Note that in addition to generating a trace to catch timing-related bugs, this provides performance analysis data Table 4.4.

Again, postprocessing tools can be used on the trace data to determine statistics such as instructions most frequently used and instructions that took the longest time. What we have so far is a real-time instruction trace with pointer information output on one external data pin, clocked at the processor core clock frequency on one clock pin. From this trace, we can determine how long each instruction takes to execute.

Normal program execution is rarely linear. Any code branching results in nonlinear code flow. Therefore, another two states must be added to determine whether a branch was executed, so the post-mortem tool can correctly calculate the new pointer from the current pointer, because it may no longer be current pointer + 4. Here's an updated example with two new states to handle whether the instruction was a branch that was taken:

In Table 4.5, IE state refers to whether an instruction executed on a cycle, and BT state refers to whether an instruction was a branch that was taken on that cycle. For the most common case of branching, the branch target address can be calculated from the branch instruction encoding itself. To postprocess the address information, instruction memory or a code listing is read to determine the target address of the branch. Table 4.6 includes the branch target addresses (BTAs) for all branches that

Table 4.3 Linear instruction trace

Cycle number	State	Did instruction execute?
Cycle 1	0	Instruction did not execute
Cycle 2	1	Instruction did execute
Cycle 3	0	Instruction did not execute
Cycle 4	0	Instruction did not execute
Cycle 5	1	Instruction did execute
Cycle 6	1	Instruction did execute

Table 4.4 Linear instruction trace reconstruction

Address	Cycles per instruction	Total number of cycles
0×10	2	2
0×14	3	5
0×18	1	6

Table 4.5 Nonlinear instruction trace

Cycle number	IE State	BT State	What happened?
Cycle 1	0	0	Inst. did not execute, not a taken branch
Cycle 2	1	0	Inst. did execute, not a taken branch
Cycle 3	0	0	Inst. did not execute, not a taken branch
Cycle 4	0	0	Inst. did not execute, not a taken branch
Cycle 5	1	0	Inst. did execute, not a taken branch
Cycle 6	1	0	Inst. did execute, not a taken branch
Cycle 7	1	1	Inst. did execute, taken branch
Cycle 8	1	0	Inst. did execute, not a taken branch
Cycle 9	1	1	Inst. did execute, taken branch
Cycle 10	1	0	Inst. did execute, not a taken branch

Table 4.6 Nonlinear instruction trace with branch target addresses

Cycle	IE State	BT State	BTA	What happened?
Cycle 1	0	0	N/A	Inst. did not execute, not a taken branch
Cycle 2	1	0	N/A	Inst. did execute, not a taken branch
Cycle 3	0	0	N/A	Inst. did not execute, not a taken branch
Cycle 4	0	0	N/A	Inst. did not execute, not a taken branch
Cycle 5	1	0	N/A	Inst. did execute, not a taken branch
Cycle 6	1	0	N/A	Inst. did execute, not a taken branch
Cycle 7	1	1	0×24	Inst. did execute, taken branch
Cycle 8	1	0	N/A	Inst. did execute, not a taken branch
Cycle 9	1	1	0×04	Inst. did execute, taken branch
Cycle 10	1	0	N/A	Inst. did execute, not a taken branch

Table 4.7 Nonlinear instruction trace reconstruction

Address	Cycles per instruction	Total number of cycles
0×10	2	2
0×14	3	5
0×18	1	6
$0 \times 1C$	1	7
0×24	1	8
0×28	1	9
0×04	1	10

have been taken. Remember, the branch target addresses (0×24 and 0×4 in this example) are not broadcast over the trace pins, but rather are determined from the debug tool after the trace is run by either reading instruction memory or reading a static code listing.

Now we can show the postprocessing for this trace. Keep in mind that the instructions that were executed at cycles 2, 5, 6, 8, and 10 may be conditional branch instructions, but if they were, the conditions to take the branch were not met (Table 4.7).

The code flow is no longer linear; the instruction at address 0×20 has not been executed. Also note that the last instruction is at address 0×04, a lower address than the start of the trace. To total up the pin count, we are using one pin for clocking and two others to handle the four finite states (one of which, executing a branch when an instruction has not been executed, will never occur), for a total of three pins.

Branches are one example of nonlinear code execution. But so far we have only handled one type of branch, one whose target address can be calculated simply by knowing the branch instruction. There are other kinds of branch instructions in which the branch target address can only be determined by a value in a designated register. These are handled as a special case in the same class as tracing interrupts.

Interrupts present a unique problem in that when an interrupt is taken, the next instruction address may jump to any one of a number of possible locations depending on the type of interrupt. These address locations are known as exception vectors. If pin bandwidth is used to create a state for every possible interrupt, the cost benefit of low pin count will be lost. Therefore, instead of tracing the type of interrupt, the address of the exception vector is broadcast. For a 32-bit RISC machine, instruction addresses are 32 bits in length. But the two least significant bits are not needed because they must always be zero, as instruction lengths are 4 bytes (32 bits). To broadcast the important 30 bits of address, only four pins are used – one pin to indicate if an address is being broadcast, and three pins to broadcast the address in a serial fashion over 10 cycles.

For example, let us assume the processor takes an exception when trying to execute an illegal instruction. When a program exception occurs, the pointer jumps to an address with an exception vector offset of 0×0700. To illustrate, we will assume the IP after the interrupt is 0×12340700. To broadcast this address,

Table 4.8 Trace of address broadcasting pins

Cycle	Addr BC	A0	A1	A2
Cycle n - 1	0	N/A	N/A	N/A
Cycle n	1	0	0	0
Cycle n + 1	1	0	0	0
Cycle n + 2	1	1	1	1
Cycle n + 3	1	0	0	0
Cycle n + 4	1	0	0	0
Cycle n + 5	1	0	1	0
Cycle n + 6	1	0	1	1
Cycle n + 7	1	1	0	0
Cycle n + 8	1	1	0	0
Cycle n + 9	1	0	0	0

the two least significant bits are ignored, because they are always zero, and the resultant octal number is 0443200700. Assuming we broadcast the least significant bits first, an address broadcast portion of a trace is shown in Table 4.8 (starting at cycle n, with A0 as the most significant bit and A2 as the least significant bit). Although not shown in Table 4.8, the IE state and BT state information would continue to be output in parallel with any address broadcast. The objection to be raised at this point is how 10 cycles can be dedicated to broadcasting the address information without ever slowing down the processor. In situations that require address broadcasting fewer than 10 cycles apart, how can the trace (and the controller) continue to run at full speed? The answer is that there is enough on-chip buffering of address broadcast information to be confident that for any realistic code sequence, processor execution will not be halted. This address broadcast mechanism is also used to handle the special branch instructions not previously considered.

4.3 Hunting Code Errors with Self-Trace

Hunting code errors with self-trace is a key issue for software developers' code analysis. Instrument trace allows trace data to be sent to a debugger host for offline evaluation. At high processor speeds, instructions need to be filtered, because one can send only a limited amount of data per clock cycle. Often the information one need was not collected or was pushed out the back of the buffer if the buffer is not "infinite" (i.e. a storage device). To find your bug, one need a specialized trace.

Specialized trace peripherals buffer certain types of useful information. For example, a discontinuity trace will track the most recent branches, as well as provide an accurate measure of the number of cycles actually used, reflecting cache and pipeline efficiency. Tracking the gross movements of the program counter enables one to trace code execution using much less information than a full instruction trace requires. If one find the program counter in a place it should not be, one can see where the code veered off.

Another useful technique is tracking jumps to uninitialized memory. First, write NOPs throughout uninitialized blocks of memory, and set the final instruction word as a breakpoint. In this way, a branch to any part of uninitialized memory will fall through to the breakpoint. One can then look back through the discontinuity buffer to discover the errant jump. Consider leaving this capability enabled in deployed devices. When the breakpoint is executed, write the specialized trace buffers and any other important system variables to nonvolatile memory. One will then have a record of invaluable debug information for hunting down intermittent bugs.

Chapter 5
An On-Chip Debug System

In this chapter we look at an on-chip debug system (OCDS) that addresses processor instrumentation requirements through a JTAG interface. The general on-chip debug system discussed provides a range of typical hardware monitoring and debug control features for a design. Notably it allows several breakpoints to be set and memory locations to be observed during run time. This example does not provide trace capabilities; they are discussed in a separate chapter.

Figure 5.1 is an example of the modular debug interface discussed in previous chapters. The JTAG interface is constructed as a separate module, with the debug port and OCDS similarly modular.

The overall OCDS consists of three blocks:

- The OCDS module (including debug control, access and triggering subsections).
- The core debug port.
- The JTAG module.

The purpose of OCDS is to debug processor operations in a systems environment. In order to do this, the OCDS should provide several capabilities, in terms of control, triggering, and information capture.

Breakpoints provide the most common means of controlling debug functions. Breakpoints operate by comparing hardware, software, or external pin signals to a predefined value and triggering events. Breakpoints have come to be referenced as generic terms for any triggering operation, particularly for a processor, that supports debug actions. In a more precise definition, breakpoints refer to triggers that break the sequential operation of a processor, by putting the processor in either a halt mode, where more exhaustive analysis can be performed while the systems are in a steady-state condition, or a stall mode, where the instruction sequencing can be manually controlled by controlling and monitoring the program counter (PC).

Other types of trigger operations that do not result in the processor halting are also referred to as breakpoints, but they should more exactly be referred to as watchpoints (whether an internal or external status or flag may be set upon triggering) or tracepoints (where a trace operation is performed (either starting or stopping a sequential trace or taking a trace snapshot)) when a trigger occurs. Watchpoints and tracepoints typically do not involve the processor operations stopping when they occur.

N. Stollon, *On-Chip Instrumentation: Design and Debug for Systems on Chip*,
DOI 10.1007/978-1-4419-7563-8_5, © Springer Science+Business Media, LLC 2011

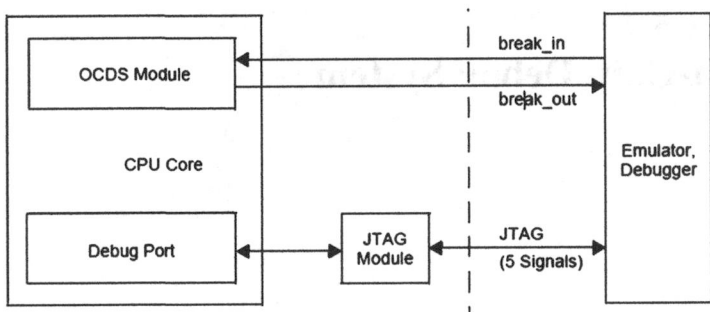

Fig. 5.1 OCDS overview

Perhaps the most typical point to perform a trigger operation is on the instruction pointer of a processor. By setting a trigger against the processor instruction value, the user can determine and track the occurrence of an instruction in the processor flow. This type of analysis is useful for tracking conditional instructions to determine if the processor has branched or jumped based on some system condition or data value.

A processor OCDS typically has multiple instruction pointer breakpoints to allow monitoring of several instructions concurrently. To evaluate instructions, it is also often required to compare against hardware values. These values may be data values that the instruction is processing or register values that are set by an instruction.

Data values often include the ability to compare only a portion of the data using masks. Using masks dramatically increases the flexibility of a data comparison, allowing a range of values to be compared without having to include much more expensive (both logic and timing) generic range logic.

Similarly, masks can be used to compare against a single bit in a status or flag register for triggering purposes. Loading compare values may be done in a variety of ways. When using external debugger software, trigger and control registers may be loaded via JTAG. Alternatively, if the registers are memory mapped, as is typical for most modern processors, trigger information may be loaded via a monitor program by the processor itself.

5.1 OCDS Features

The OCDS, independently from the JTAG logic, may support other interfaces and underlying protocols, which may include packet or direct parameter loading. Packets provide a pre-defined mechanism for loading and updating information between an off-chip controller and a target. Standards such as Nexus 5001, which is discussed in Chap. 11, use a packet-based protocol. Other debug systems, notably those used by processor vendors, use a direct parameter-loading method, where individual debug parameters and register values are loaded on an individual basis to allow finer granularity and marginally better efficiency when compared to packet-based transfers.

JTAG interfaces support both approaches because they basically involve register loads of different lengths. An additional function supported by many OCDS is the ability to emulate read/write functionality. On-chip memory may be accessed by the debug port in a variety of ways, including direct access to the memory block to load and store addresses and data, access to the memory bus via the memory controller, or the OCDS taking control of the on-chip bus directly (becoming a bus master) to generate memory access.

Each of these approaches has relative benefits. In direct access, the debugger has the lowest granularity of control. Basically, the debugger presents the external address and enable information to the memory block and either loads external data into the memory or reads out memory data for an address through the debug port. Depending on levels of hierarchy in the memory and types of access methods, this may require several clock cycles to download the memory inputs to the chip and then to export memory data as needed. Although this has a fair amount of overhead, direct access to the memory allows the most flexibility, including the ability to operate on a memory in a processor that is stalled, in reset, or powered down.

Unless a significant amount of DMA functionality is added to the debug block, the limitations of atomically accessing each memory and location can be limiting to access or large or multiple memories. An alternative method is to have the debug block access or control the DMA or memory controller logic for a memory, by sending a command to the memory controller to prioritize and support debug interface transfers The advantages of this are the ability to use the resources of the memory controller to provide bursting or sequential access to memory data without having to directly control every access of the memory.

A disadvantage of both direct access and DMA/memory control is that memory, as a shared resource, is not always available for debug operations. In some systems, the OCDS may take bus mastership of the systems to have uninterrupted control for memory access. The advantage of this is increased access to all memories and memory-mapped registers of a system. As a bus master, the debug interface may also be able to trigger memory dumps through an external bus. The slip side of a debug bus master is increased complexity of the memory interface, due to bus control logic as part of the OCDS and increased complexity in the bus fabric itself to accommodate the debug bus master, and the possibility of intrusive impact to other parts of the system. This impact is reduced in modern bus systems where interfaces are point to point within a bus fabric.

Tampering through the debug interface is increasingly a concern in modern systems. It is increasingly realized that an OCDS supporting memory and on-chip data access is also a potential vulnerability in allowing hackers access to this memory. An OCDS may have security mechanisms and management, separate and independent of any security features of the JTAG interface, in order to protect it from unauthorized access.

A summary of OCDS features are:

- Support for communication between monitor and external debugger.
- Optional error protection.
- A security mechanism to allow authorized access only.
- Low-end tracing through reads (writes) triggered by the OCDS.
- Fast tracing through transfer to an external bus.
- Analysis registers for internal bus locking situations.
- Several OCDS can be operated across a single JTAG interface.
- Control and data transfer mechanism for OCDS.
- A data transfer channel for programming on- and off-chip (nonvolatile) memory.
- An access port for on- and off-chip (across external bus controller) system analysis and configuration.
- A data channel that is independent of user resources.
- An API that allows multicore debugging.

The target application of the OCDS is use of the JTAG interface as an independent port for OCDS. The external debug hardware can access the OCDS registers and arbitrary memory locations. Multiple OCDS may be operated through a single JTAG (or other) interface to provide a more effective debug solution for multicore debugging modules operated from standard debuggers in one debug session. The JTAG API provides a straightforward proven interface for standard debuggers and arbitrates access of the JTAG interface in a transparent way.

In order to protect the system during normal execution, the OCDS is typically disabled by default. Events can be generated only when the OCDS is enabled. The OCDS module has an enable signal that is normally connected to the chip's internal JTAG reset. This means that the OCDS is enabled when the JTAG module is not in the reset state. This is usually the case when the external debugger is connected.

Depending on the system architecture, the enable signal may be controlled by another source, or the OCDS module can be optionally enabled by software. The debug concept addresses both the generation of debug events and the definition of event actions taken when a debug event is generated.

5.1.1 Debug Events

- Hardware trigger combination.
- Execution of a DEBUG instruction
- Break pin input.

5.1.2 Debug Event Actions

- Halt the processor.
- Call a monitor.

- Trigger a data transfer (TRIGGER DATA TRANSFER).
- Activate an external pin.

5.1.3 Debug Registers

The key to any OCDS is a defined register set that supports different operations. Many of these will be discussed later in the chapter. These typically include:

- A debug status register that contains the status.
- A debug instruction pointer that contains a pointer value.
- A debug activity register that specifies an action if a DEBUG instruction is executed.
- A debug external event register that specifies an action if an external break pin is asserted.
- A debug hardware trigger register that specifies hardware triggers and action.
- A debug data programming register that triggers to the data programming register for debug hardware compare.

The debug select and programming register for debug hardware compare and debug task ID register is used by advanced real-time operating systems to store the task ID of the active task, which contains:

- a 24-bit instruction pointer.
- R_ADR 24-bit data address of reads.
- W_ADR 24-bit data address of writes.
- DA 16-bit data value (reads or writes).

Debug hardware event equals comparison register 0,
Debug hardware event equals comparison register 1.
Debug hardware event equals comparison register 2.
Debug hardware event range comparison register (greater).
Debug hardware event range comparison register (less).

The trigger sources (discussed later) are compared and combined in the hardware trigger generation unit (see Fig. 5.8). The hardware trigger generation unit is programmable with the debug event control register and consists of two paths. The upper path is for one range comparison and the lower path for three equal comparisons. The equal path can be optionally configured for two masked equal comparisons.

5.2 Operation Modes

OCDS can be used for two different purposes. The first is to read and write memory locations (RW mode) and the second is to exchange data with a program (monitor) running on the processor (communication mode).

RW mode is used by the external debugger host to read or write memory locations. In RW mode, the instructions IO_READ_WORD, IO_WRITE_WORD, IO_READ_BLOCK, IO_WRITE_BLOCK, and IO_WRITE_BYTE are used in their generic meaning. The data address is in IOADDR and is set with IO_SET_ ADDRESS. RW mode needs the TRIGGER DATA TRANSFER interface to actively request data reads or writes.

The default data type is a 16-bit word, used for single-word transfers and block transfers. If the external debugger host wants to read a single byte, it must read the associated word (IO_READ_WORD) and extract the needed byte. Writes to bytes are supported with the IO_WRITE_BYTE instruction. In addition, for this instruction, the external debugger host must shift in the full word, but only the selected byte is actually written. The position is defined by the lowest address bit in IOADDR.

The TRIGGER DATA TRANSFER interface does the actual read or write of memory locations. It is configured with transactions requested by the JTAG shift core. The data is transferred to/from the RWDATA register. TRIGGER DATA TRANSFERs typically have the highest processor priority.

Communication mode allows communication between an external debugger host and a program (monitor) running on the processor. In this mode, the external debugger host is master of all transactions. The external debugger host requests the monitor to write or read a value to/from COMDATA. One difference from the RW mode operation is that in communication mode, the read or write requests are not actively executed by OCDS, but it sets request bits in a processor-accessible register to signal the monitor that the debugger host wants to send (IO_WRITE_WORD) or receive (IO_READ_WORD) a value. The monitor must poll this I/O status register (IOSR). The IOADDR register is not used. The debugger host and monitor exchange data directly with the COMDATA register. Communication mode ensures that all send and receive transactions are served under all conditions in the correct sequence, even if the OCDS changes to RW mode.

5.2.1   Entering Communication Mode

Communication mode is the default mode after reset. If OCDS is in RW mode, communication mode is entered when the external debugger host writes to the MODE bit in the IOCONF register.

5.2.2   Communication Mode Instructions

Communication mode uses only the IO_WRITE_WORD and IO_READ_WORD instructions. An IO_SET_ADDRESS instruction sets IOADDR just as in RW mode (no effect for communication mode).

5.2.3 Monitor-to-Debugger Host Data Transfer (Receive)

The CRSYNC bit signals the monitor (processor) that the external debugger host wants to receive a new COMDATA value. It is set in communication mode with the active *read request* signal for the IO_READ_WORD instruction. The CRSYNC bit is automatically cleared when the monitor (processor) writes to COMDATA independent of the mode (communication mode or RW mode). The debugger host can request data, do something in RW mode, and then fetch the requested data with the next receive cycle.

5.2.4 Debugger Host-to-Monitor Data Transfer (Send)

The CWSYNC bit signals the monitor (processor) that the external debugger host has written a new value to the COMDATA register. It is set in communication mode with the IO_WRITE_WORD instruction. The CWSYNC bit is cleared when the monitor (in the processor) sets an acknowledge bit in IOSR independent of the mode (communication mode or RW mode). This allows sending data in communication mode, switching to RW mode, and then performing other operations without having to wait until the monitor has read COMDATA. The next time that communication mode is entered, busy bits are output when COMDATA was not already read by the monitor.

Note that in the case of a send (IO_WRITE_WORD) followed by receive (IO_READ_WORD), both bits CWSYNC and CRSYNC are set and must be served by the monitor in this sequence. A previous receive request blocks the send. This means that a requested value must be fetched by the debugger host before it issues a new send command.

5.2.5 High-Level Synchronization

To improve the robustness of the communication channel, it is helpful to distinguish between commands from the debugger and regular data exchange. For example, if the debugger aborts its request just when the monitor responds, the high-level synchronization between the debugger host and the monitor would be lost.

To prevent this, a COM_SYNC bit is provided to synchronize the communication channel between the debugger and the monitor on a higher level. It is set in the IOCONF register and can be read in IOSR by the debugger. The debugger/monitor can simply use this bit to reset the communication channel or, for more advanced use, this bit can tag data from the debugger to the monitor as instructions.

5.3   OCDS Registers

Debug status register contains several types of information about the current status of the OCDS, including:

- It indicates whether the debug support is enabled.
- It gives the source of the last debug event.
- It gives the system debug state.

Key fields for the status register include:

- DEBUG_STATE: The current debug state is user mode, software debug mode, or halt debug mode.
- OCDS_P_SUSPEND: This causes sensitive peripherals to suspend operation by controlling a peripheral suspend signal. If set, all sensitive peripherals will suspend. This bit is set by a debug event according to the associated PERIPHERALS_STOP bit in the active debug event control register. This bit must be reset by the debugger.
- TRGEVT_R_CMP: This is a comparison matched for the current event.
- TRGEVT_E_CMP0: This is a comparison matched for the current event.
- TRGEVT_E_CMP1: This is a comparison matched for the current event.
- TRGEVT_E_CMP2: This is a comparison matched for the current event.
- EVENT_SOURCE: This reports the source of the last debug event, which is one of the following:

1. External break pin (debug hardware trigger).
2. Debug instruction executed (debug external event).
3. Hardware trigger combination (debug data programming).

5.3.1   Debug Task ID Register

TASKID is an input to the hardware trigger event generation unit intended to be used by advanced real-time operating systems to store the task ID of the active task.

5.3.2   Instruction Pointer Register

This register makes the instruction pointer visible when the processor is in halt mode.

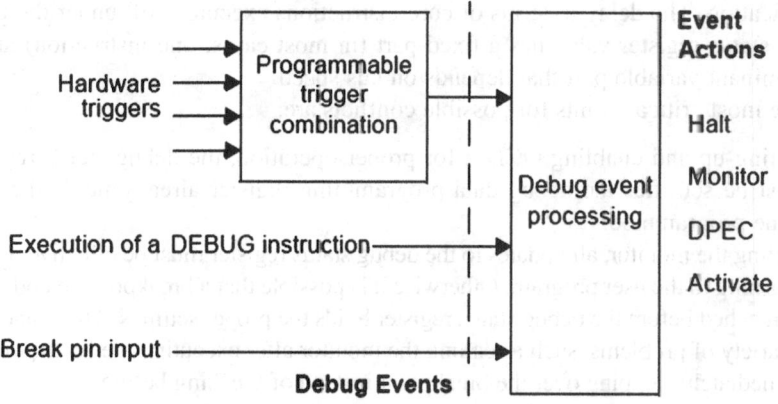

Fig. 5.2 OCDS module block diagram

5.3.3 Hardware Trigger Comparison Registers

The DEBUG HARDWARE COMPAREn registers are used in the hardware trigger event generation unit as reference values for the comparisons (Fig. 5.2). They can be programmed with special function registers, and the selected comparison register compares information as discussed in the next section:
Select DEBUG HARDWARE COMPARE0
Select DEBUG HARDWARE COMPARE1
Select DEBUG HARDWARE COMPARE2
Select DEBUG HARDWARE COMPAREL
Select DEBUG HARDWARE COMPAREG
DEBUG INSTRUCTION POINTER is the current instruction pointer in halt mode and is valid in halt mode only.

5.3.4 Considerations on Accessing OCDS Registers

The functions of OCDS are generally controlled by writing to the debug status register. To be executed correctly, any debug step needs the respective bit fields being used to have new values effective (this depends on the speed of the bus). This becomes more important as the bus speed becomes lower compared to the core speed, that is, compared to the speed of executing instructions. For a pipelined machine, different read/write operations may be executed at different pipeline-stages. A basic potential problem to be kept in mind is the new debug status register value cannot as a rule be effective for the instruction immediately following its modification. The delay in terms of core instructions executed still under the prior debug status register value has a fixed part (in most cases, one instruction) and a predominant variable part that depends on bus speed.

The most critical points for possible conflicts are:

- Setting-up and enabling OCDS; for proper operation, the debug status register must be set after the debug data programming register already holds the new value programmed.
- Exiting the monitor, all updates to the debug status register must be effective before returning to the user program. Otherwise it is possible that a breakpoint in code will be reached before the debug status register holds the proper settings. This can cause a variety of problems, such as calling the monitor after executing the breakpoint or immediately stepping over the breakpoint instead of breaking before it.

The principal solution to avoid problems accessing OCDS registers is to ensure that after an instruction writes to a register, the instruction that uses the new value will be executed only when the new settings are really effective.

Use noncritical instructions after writing to an OCDS control register (i.e., debug status register); instructions should be used in which execution does not depend on the new settings, so it is sure that the new debug status register value is effective before continuing with the next instruction. This is independent of the bus speed because the processor ensures the write operation is completed, before continuing with the next read from the same location. Consequently, this is the easiest and most reliable decision to ensure proper OCDS operation.

If the OCDS is disabled (usually when the JTAG module is in the reset state), the OCDS module and all its registers are reset with every processor reset; otherwise, it is never reset. This behavior allows a defined reset in the cases when no debugger is connected or the debugger controls the OCDS indirectly with a monitor. In the other case, when the debugger controls the OCDS directly, the OCDS registers are not affected by user, program, or system environment resets. This permits very unfriendly systems to be debugged as well.

5.4   OCDS JTAG Access

JTAG operations allow access to the JTAG module. In addition to OCDS-specific instructions, it supports standard (required) JTAG instructions and the JTAG BYPASS registers and two OCDS-specific CCONF and IOPATH registers that communicate with the OCDS logic block (Fig. 5.3 shows only the OCDS specific portions of the JTAG module).

ID register implementation is a product-specific decision. This allows maintenance of one central version and part number register that can be accessed either from the processor as an SFR or across JTAG with the IDCODE instruction. According to the JTAG standard, the IDCODE instruction must have the structure as discussed in Chap. 3.

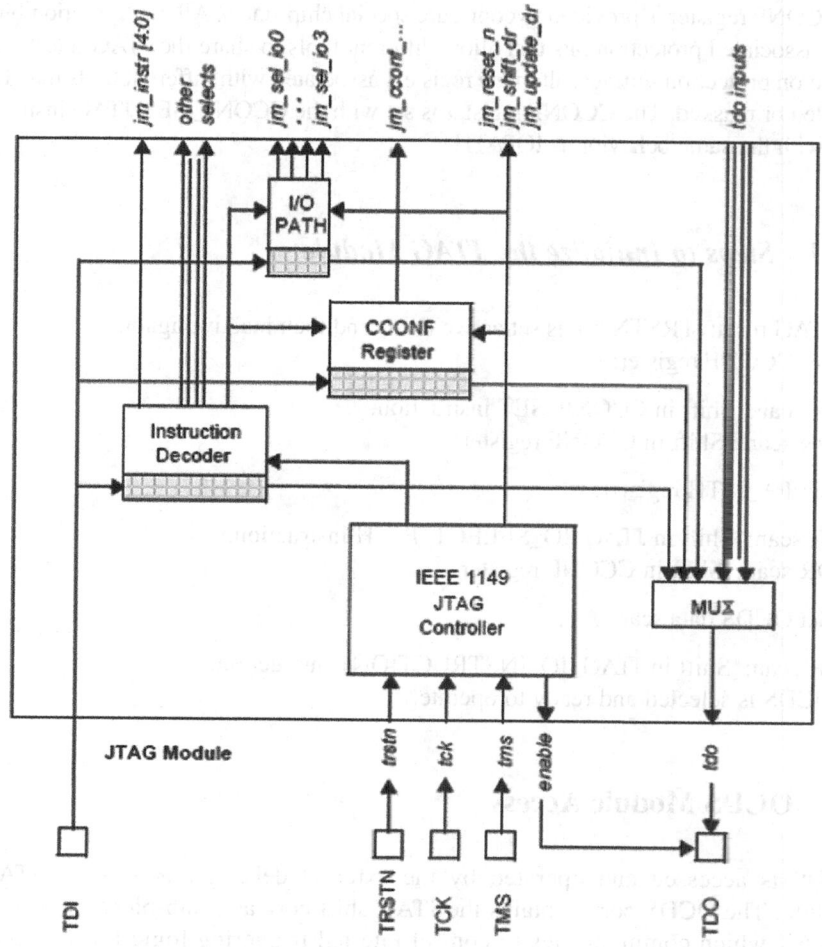

Fig. 5.3 JTAG module and interfaces to the OCDS

For the BYPASS instruction, the TDO output is equal to TDI, delayed by one TCK cycle.

IOPATH register is a modified JTAG scan register that stores a copy of the TDO to provide error protection. The TDI/TDO behavior is the same as for a JTAG BYPASS instruction except that the first bit output (state capture-DR) is 1. This difference is important if there was a bit error when the JTAG instruction was shifted in. In the most probable case, when this faulty JTAG instruction is not implemented, the JTAG module would set the BYPASS mode, which could not otherwise be distinguished from the JTAG_IO_SELECT_PATH instruction.

The IOPATH register is used to select OCDS. If the JTAG instruction is in the I/O address range, the associated select signal is active. IOPATH register is set like a regular JTAG scan chain register with the JTAG_IO_SELECT_PATH instruction.

CCONF register is provided to configure special chip states. All configuration bits have associated protection bits that allow different tools to share the JTAG interface. Based on protection settings, alternate registers associated with different tools may be enabled or masked. The CCONF register is set with the CCONF_SET JTAG instruction with the same behavior as IOPATH.

5.4.1 Steps to Initialize the JTAG Module

1. JTAG reset: TRSTN pin is set active (low) and then inactive again.
2. Set CCONF register:

   IR scan: Shift in CCONF_SET instruction.
   DR scan: Shift in CCONF register.

3. Set IO_PATH register:

   IR scan: Shift in JTAG_IO_SELECT_PATH instruction.
   DR scan: Shift in CCONF register.

4. Set OCDS data scan:

   IR-scan: Shift in JTAG_IO_INSTRUCTION1 instruction.
   OCDS is selected and ready to operate.

5.5 OCDS Module Access

OCDS is accessed and operated by the external debugger across the JTAG module. The OCDS core contains the JTAG shift core as a sub-block, shown in Fig. 5.4 which communicates to control internal triggering logic for data processor execution control (DPEC) transfers and Bus monitoring. The JTAG shift core is controlled by the JTAG signals (Fig. 5.5) and therefore is asynchronous to the rest of the OCDS core. OCDS is considered busy when the requested read or write operation has not yet been finalized. The external debugger host is master of all transactions, initiating the transfers for both directions.

5.5.1 Error Protection

The JTAG standard does not include any error protection for serial transmission (TDI and TDO pins) and control (TMS pin). However, there are some ways to include error protection without extending too much beyond the JTAG framework.

Error protection for input data (TDI) is achieved by making input data directly observable on the output pin (TDO) with one clock cycle delay. Output data can be shifted out twice (multiple) and then compared for maximum error protection.

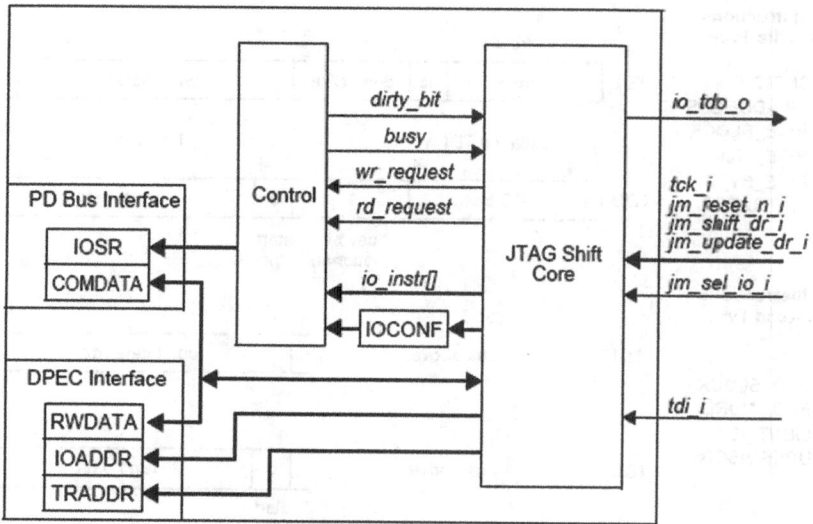

Fig. 5.4 OCDS module and JTAG interfaces

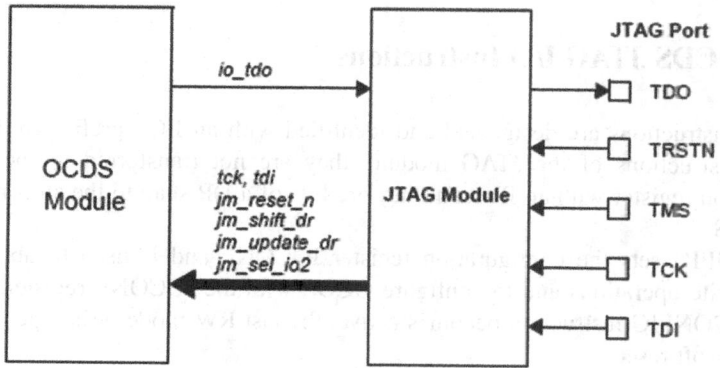

Fig. 5.5 OCDS and JTAG module connections

When OCDS is selected, it is controlled with the TDI bit stream with the JTAG sequence: Capture_DR, multiple Shift_DRs, and Update_DR. The first four bits shifted in are the I/O instruction. The next bits (busy bits) are ignored, until a start bit occurs on TDO. Busy bits can occur for all I/O instructions except IO_CONFIG, when the previous operation has not yet finished, as shown in Fig. 5.6.

If the instruction is a write-type instruction, the TDI bit, in parallel to the start bit, is used as the first data bit, followed by the rest of the data and ending with a "don't care" bit. If more data bits are shifted in than required, the first (superfluous) data bits are ignored and the last are used for the update.

If the instruction is a read-type instruction, all TDI bits after the instruction are ignored after the start bit on TDO; the read data is shifted out. If the instruction is undefined or not implemented, the client responds with an indefinite number of busy bits.

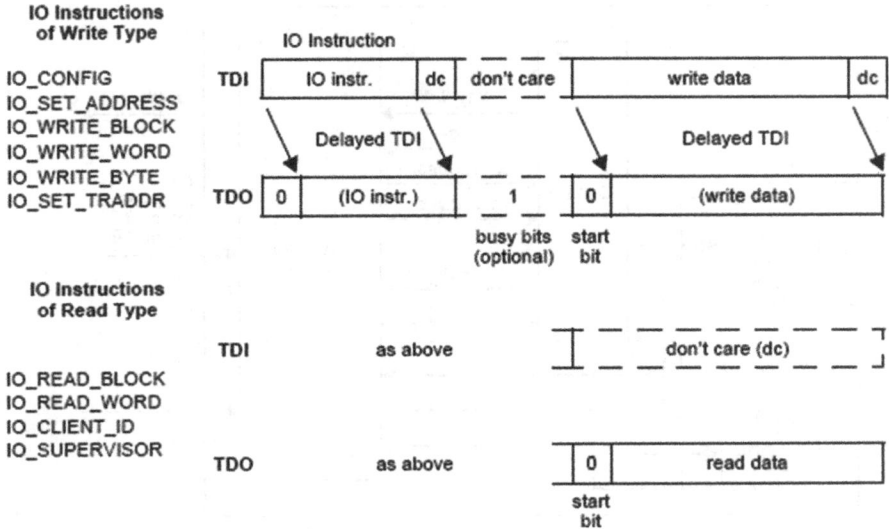

Fig. 5.6 Serial TDI and TDO transfers in Shift_DR state

5.6 OCDS JTAG I/O Instructions

OCDS instructions are designated and identified with an I/O_ prefix. Unlike the JTAG instructions of the JTAG module, they are not transferred to the JTAG instruction register with an IR scan; they are bits of a DR scan to the shift register of OCDS.

IO_CONFIG sets the configuration register IOCONF and is used to abort RW mode write operations and to configure OCDS with the IOCONF register. When the IO_CONFIG instruction becomes active, the last RW mode write operation is aborted (soft reset).

IO_SET_ADDRESS sets the address IOADDR for the next RW mode access.

IO_READ_WORD is used to read data in RW mode or to receive data in communication mode.

IO_READ_BLOCK reads the data block starting with the address in IOADDR and is used in RW mode only. The only difference from IO_READ_WORD is that the address for IO_READ_BLOCK is post-incremented by a word address. Read instructions can be aborted when the external debugger host sets the Update_DR state. For IO_READ_WORD in communication mode, at least four shift cycles must occur after the output of the start bit to acknowledge the read. This prevents the loss of read data words.

IO_WRITE_WORD is used to write data in RW mode or to send data in communication mode.

IO_WRITE_BLOCK writes to the data block starting with the address in IOADDR and is used in RW mode only. The only difference from IO_WRITE_WORD is that the address for IO_WRITE_BLOCK is post-incremented by a word address. For multiple write instructions, enough shift cycles must occur after the output of the start bit for the write from the Update_DR state to allow the last write to be checked before initiating a new write.

IO_WRITE_BYTE is a special case of IO_WRITE_WORD for writing bytes. For IO_WRITE_BYTE, it is required that a complete 16-bit word must be shifted in from which the lower byte is always written (for even and uneven addresses).

IO_SET_TRADDR sets the TRADDR register, which is used for tracing with an external bus address.

IO_SUPERVISOR is used to release RW mode and communication mode from the error state. This instruction also outputs the IOINFO register after a start bit.

IO_CLIENT_ID returns a client-specific ID code from register CLIENT_ID.

IO_SET_TRADDR sets the TRADDR register.

IO_SUPERVISOR acknowledges reset and analyzes bus-locking situations.

IO_CLIENT_ID reads the Client ID.

Figure 5.7 shows the relationships among TDI, TDO, and the shift register content after the client instruction has been shifted in. MUX1 is controlled by the active instruction, MUX2 is controlled by the status of the client (busy or operation finished).

In the case of I/O write-type instructions, after the TDO start bit occurs, the delayed data is shifted into the shift register and in parallel is output on TDO. In the case of I/O read-type instructions, the captured data is shifted out via MUX1 and MUX2.

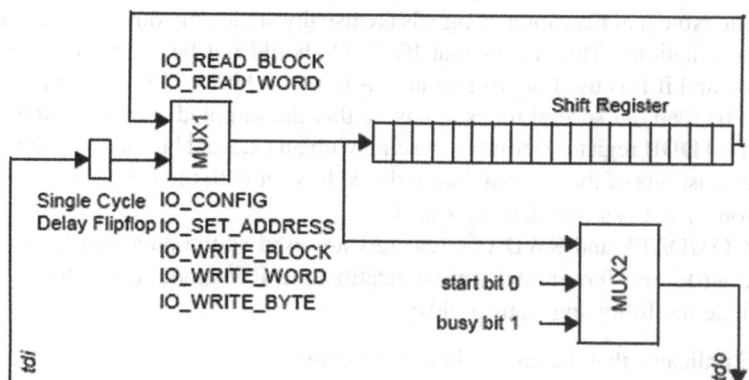

Fig. 5.7 Shift register behavior in the Shift_DR state

5.7 OCDS JTAG Registers

These are registers internal to the OCDS. Unless otherwise noted, they are externally accessed using the OCDS instructions.
CLIENT_ID allows that the external debugger checks the hardware in an auto-configuration mode and includes the following fields:

- IOADDR is the address for the next RW mode access.
- IOCONF is the configuration register.
- IOINFO is the chip state analysis register.
- TRADDR is the external bus trace mode address.
- COMDATA is the communication mode data register.
- RWDATA is the RW mode data register.
- IOSR is the status register.

IOADDR register holds the 24-bit address for the next OCDS access. IOADDR is updated in the Update_DR state with the shift register contents when the IO_SET_ADDRESS instruction is active or incremented by two (a 16-bit word) if an IO_READ_BLOCK or IO_WRITE_BLOCK instruction has been executed.
IOCONF register is used to configure OCDS. The IOCONF register is write only for the debugger host and is not accessible from the processor side.

- The MODE bit determines whether OCDS is in RW or communication mode.
- TRIGGER_ENABLE enables triggered transfers in RW mode. The next transfers must be triggered by the TRIGGER DATA TRANSFER event action provided by the OCDS module.
- The EX_BUS_TRACE bit enables triggered transfers to an external bus address.

The IOINFO register is provided to analyze bus locking situations or certain other chip internal error states. It is not a physical register, but it represents certain chip state information. After an IO_SUPERVISOR instruction, this information is shifted out. Note that the captured signals are usually static only during these locking and error situations. This means that IOINFO should not be used during normal operation, and if it is used in error situations (no start bit for RW mode operation), it should be read out several times to ensure that the sampled values are static.

The TRADDR register is used for tracing with an external bus address. It defines the uppermost bits of the external bus address. It is set with the IO_SET_TRADDR instruction by the external debugger host.

The COMDATA and RWDATA registers are used as the data register for both read and write transfers in and communication and RW mode, respectively. They also include the following status fields:

- IDLE indicates that the chip is in an idle state.
- POWER_DOWN indicates that the chip is in the power down state.
- EXTBUS_HOLD indicates that the exterior bus is busy.
- LMBUS_HOLD indicates that the local memory bus is busy.
- PBUS_HOLD indicates that the peripheral bus is busy.

The IOSR register is used in communication mode to disable OCDS from the processor side for security reasons and to perform monitor-controlled. The IOSR register is only accessible from the processor side and includes the following fields:

- RW_DISABLE is used to prevent OCDS from entering RW mode. It can only be set by the processor in communication mode. If OCDS has already entered RW mode, all attempts by the processor to set this bit are ignored.
- RW_ENABLED is provided to enable the user program to store whether RW mode is enabled already.
- DBG_ON indicates whether an external debugger is present.
- CLNT_ON indicates whether the OCDS is currently selected by the external debugger. It is directly controlled by the OCDS select signal that is set with the IOPATH register in the JTAG module.
- MTR_CTRL is a monitor-controlled tracing field that can be used by a monitor to control the tracing of memory locations. Note that this feature may be used only if no external debugger controls OCDS across the JTAG interface.

5.8 Hardware Triggers

A triggered event may occur due to either trigger operations occurring in the OCDS or external debug break pins allowing the debugger to asynchronously interrupt the processor. The action taken when this signal is asserted for debug hardware trigger operation depends on the debug control registers.

It is possible that more than one event may be raised in a single cycle. In this case, the priority of events to be handled is usually based on the sequence in which the events appear in the event sources list; those listed first are handled before those listed later.

Different events will have different priorities; typically break operations have a priority of:

1. Pin input debug hardware trigger (highest).
3. Execution of a debug instruction debug external event.
3. Hardware trigger combination.
4. Debug data programming.

Hardware trigger fields allow range comparison input on the following:

1. Instruction pointer (IP).
2. Data value (DA).
3. Write address (W_ADR).
4. Read address (R_ADR).
5. Equal compare input MUX control (see Fig. 5.8).
6. Instruction pointer (IP).
7. Data value (DA).
8. Write address (W_ADR).
9. Task ID (TASKID).

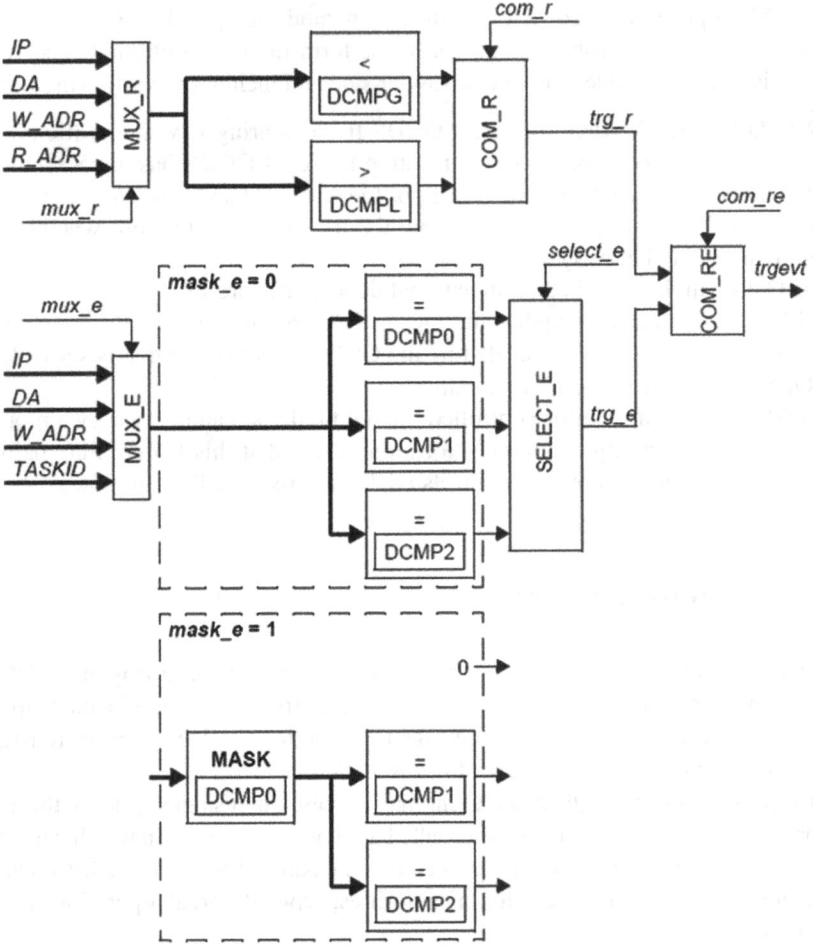

Fig. 5.8 Hardware trigger generator

Hardware triggers can enable the following debug related operations:

1. *Execution of a DEBUG instruction.* In many processors, there is a mechanism through which software can explicitly generate a debug event. This can be used, for instance, by a debugger to patch code held in RAM in order to implement breakpoints. A special DEBUG instruction is defined that is in the user mode instruction, and its operation is dependent on whether OCDS is enabled. If OCDS is enabled, the DEBUG instruction causes a debug event to be raised and the action specified in an external event control register is taken. If OCDS is not enabled, the DEBUG instruction may be treated as a NOP.
2. *Trigger data transfer (TRIGGER DATA TRANSFER).* Triggering the OCDS to execute a pending transfer is one of the actions that can be specified to occur when a debug event is raised. This can be used in critical routines in which the

system cannot be interrupted to transfer a memory location to the data register and read it (trace) through the debug port.

3. *Call a monitor.* Calling a monitor with a special debug hardware trap is one of the possible actions to be taken when a debug event is raised. This trap has the high priority, but the monitor routine can reduce its own priority level. This short entry to an interruptible monitor allows a flexible debug environment to be defined that is capable of satisfying many of the requirements for efficient debugging of a real-time system. For example, safety critical code can be served while the debugger is active. The monitor is ended with a regular RETI instruction. The debug flag bit DEBTRAP has to be cleared on exiting the TRAP routine; otherwise it will be called again.

5.8.1   Structure of a Noninterruptible Monitor Routine

1. Perform processing (noninterruptible).
2. Set DEBUG STATUS REGISTER.
3. Clear the DEBTRAP bit.
4. Return to THE user program with the RETI instruction.

5.8.2   Structure of an Interruptible Monitor Routine

1. Set the DEBUG STATUS REGISTER.DEBUG_STATE (user mode).
2. Clear the DEBTRAP bit.
3. Reduce the interrupt level ILVL in W.
4. Do processing.
5. Set the DEBUG STATUS REGISTER.
6. Return to the user program with the RETI instruction.

Reduction of the interrupt priority of the monitor can cause stack overflows. If the task that causes the debug event has a higher priority than the monitor, the monitor will be pushed onto the stack again and again.

Note: Care must be taken that the monitor does not cause an event itself. Otherwise it will be started again and again and cause stack overflows.

4. *Enter halt mode.* The system suspends execution by halting the instruction flow and will not respond to any interrupts. It then relies on the external debug system to interrogate the target entirely by reading and updating through the OCDS debug port.
5. *Activate an external pin.* An external pin can be controlled as a debug event action. This is to be used in critical routines in which the system cannot be interrupted to signal to the external world that a particular event has happened.

This feature could also be useful to synchronize the internal and external debug hardware or for profiling. In most cases the break out pin is active for as long as the trigger condition is met.

6. *Enable single stepping.* Single stepping can be done in halt mode or with a debug monitor.

- *Single stepping in halt mode.* For this behavior, the trigger condition is set as always true. After every restart, the processor will be halted again when the next instruction has been executed.
- *Single stepping with a debug monitor.* Single stepping can serve high-priority interrupt requests. The basic approach is similar to the single stepping in halt mode with two differences:

  – The event action is set to call a monitor.
  – The code of the interrupt service routines and of the debug monitor may not be part of the IP address trigger range.

5.8.3 Debug Event Control Registers

Each possible source of a debug event has an associated register that defines which action should be taken when that debug event is raised. The debug event control registers have the same structure for all currently defined sources.

EVENT_ACTION specifies what happens when the associated debug event is raised:

- Halt debug mode halts the processor.
- Software debug mode calls a monitor.
- Trigger a data transfer (execute TRIGGER DATA TRANSFER).
- Activate an external pin.
- Set the event in the DEBUG STATUS REGISTER.

For software and halt mode, the EVENT_ACTION sets the DEBUG_STATE field in the debug status register.

PERIPHERALS_STOP. Sensitive peripherals suspend operation if this event occurs.

ACTIVATE_PIN activates an external pin that is normally (not during debug) inactive.

The COM_R field enables the range comparison. For in-range comparisons, DEBUG HARDWARE COMPAREG is used as the upper boundary and DEBUG HARDWARE COMPAREL is the lower boundary. For out-of-range comparisons, it is the is reversed. This allows range comparison:

- In range: if debug hardware compare_greater > input > debug hardware compare_lower, otherwise 0.
- Out of range: if (debug hardware compare_greater > input) or (input > debug hardware compare_low), otherwise 0.

The MASK_E field selects unmasked or masked equal comparison and distinguishes between masked and unmasked input for the equal comparison. In the masked case, DEBUG HARDWARE COMPARE0 controls the relevant bits for the comparison.

The SELECT_E field enables the equal comparisons to be included in the trigger event generation and selects which is used, as follows:

- If debug hardware compare0 matches, otherwise 0.
- If debug hardware compare0 or debug hardware compare1 match, otherwise 0.
- If debug hardware compare0 or debug hardware compare1 or debug hardware compare2 match, otherwise 0.
- 0 (always).
- If debug hardware compare1 matches, otherwise 0.
- If debug hardware compare1 or debug hardware compare2 match, otherwise 0.

The COM_RE field selects equal and range comparison combination of either:

- The debug trigger event signal is Trigger_range OR trigger_event.
- The debug trigger event signal is _range AND trigger_event.

5.9   Additional Features

Triggered transfers (TRIGGER DATA TRANSFER) can be used to read or write a certain memory location when an OCDS trigger becomes active. Triggered transfers are executed when OCDS is in RW mode, the TRIGGER_ENABLE bit in IOCONF is active, the JTAG shift core has requested a transaction, and an OCDS TRIGGER DATA TRANSFER event occurs. Triggered transfers behave like normal transfers, except that there must also be a transfer trigger after the JTAG shift core requests the transfer.

Tracing of memory locations is one of the main applications for triggered transfers. Trace of certain memory locations can be performed when the OCDS core activates the TRIGGER DATA TRANSFER event action if this memory location is written by the user program. OCDS is configured to read the location on this trigger. The maximum transfer rate that can be reached is defined as the number of instruction cycles that need to be between two processor accesses to the memory location. The instruction cycle time of the processor is the clock rate of the JTAG interface (TCK). This requires a delay of several cycles between traces, but in many cases this will be sufficient to trace static values, for instance, the task ID register. A trace delay factor of 30 cycles is the number of cycles required: 10 bits for the JTAG state machine, I/O instruction, start bit, and transfer of 16 bits for the data and 4 bits for the synchronization between the transfer trigger and the shift out.

If the trigger rate is higher, some accesses are lost. To notify the external debugger about these missed events, a dirty_bit read tag is set. This bit is appended to the read data when it is shifted out.

Tracing with external bus address is a special operating mode of the TRIGGER DATA TRANSFER interface for faster tracing. In this mode, the data is not written to RWDATA and shifted out via the JTAG port, but rather is directly written to an external bus address. The data is then captured from the external bus by the debugger ("trace box"). This kind of tracing can be enabled in communication mode only and can be used in parallel to it.

Monitor-controlled tracing allows trace when the JTAG interface is not accessible. A monitor uses this feature only when no external debugger is connected to the OCDS across JTAG. Otherwise, errors will occur because this feature can share resources with the normal modes used by the external debugger. Monitor-controlled tracing is not a security risk. Even if it is unintentionally enabled by a user program, a transfer occurs only when the OCDS triggers it.

Monitor-controlled tracing is equivalent to triggered transfers but is controlled by a monitor running on the processor. It can be used to move an arbitrary memory location on an OCDS core TRIGGER DATA TRANSFER event action. The transfer is executed when OCDS is not selected, and there is a transfer trigger. Source and target addresses are programmed with the selected address (source or target) register.

5.9.1 System Security

After reset, OCDS is in communication mode and is brought into RW mode. If the user program running on the processor sets the RST_HLT immediately after reset, there is no way from the outside to get OCDS into RW mode via the JTAG interface.

To have a protected system in the field that can be accessed by authorized users, the following solution can be used (all bits are in the IOSR register):

- The first instruction of the user program after reset disables RW mode.
- The user program checks if an external debugger is present. If not, it continues with the regular code.
- The external debugger sends a key in communication mode.
- The user program starts to accept and compare the key some time after reset and after JTAG shift in of the send request.
- If all keys are correct, the user program resets RST_HLT and sets RW_ENABLED.
- Then the user program then knows (RW_ENABLED) that OCDS has been enabled once and thus does not prevent the enabling after the next reset.

OCDS is in power-saving mode when it is not selected from the JTAG side. The only register that is always accessible and working is IOSR. If the monitor-controlled tracing mode is enabled, the required resources are functional.

5.9.2   Reset from the JTAG Side

If the internal JTAG reset becomes active, all RW mode and communication mode requests are aborted. A JTAG reset always requires a following processor reset to ensure that the JTAG shift core and the control part of OCDS are in a defined state under all conditions.

5.9.3   Reset from the Chip/Processor Side

In this case, all I/O instructions go to an error state. The external debugger host must acknowledge this state with the IO_SUPERVISOR instruction; this is done to notify the external debugger host that something unexpected may have happened and to check connectivity of the communication channels.

OCDS enters the error state on all chip internal resets (except JTAG reset). The error state can be left with the IO_SUPERVISOR instruction. Another error state occurs when the chip internal bus is blocked for TRIGGER DATA TRANSFER transfers. If this condition occurs, the IO_SUPERVISOR instruction can be used to read the IOINFO register, which provides analysis information.

Chapter 6
Bus System Debug

In SoC platform architectures, more complexity is being added at the functional interconnect level. On-chip buses and interconnect systems for integrating IP blocks into a SoC solution have become sophisticated subsystems in themselves, with multilayer, cross-bar, and network on-chip alternatives being developed by both IP vendors and integrating companies themselves. The goal in most cases is to address, within reasonable wiring and size constraints, the increasing amount of bandwidth required for complex applications and optimized communication between different blocks of IP, both with each other and with shared resources such as memory and peripherals. In many cases these interconnecting architectures contain sophisticated internal complexity and have tunable parameters to allow trade-offs and optimization of the interconnect features for a given architecture and application. The interfaces to these interconnect systems are typically implemented at a socket level using one of several bus interface standards (OCP, ABMA AHB, and AXI being among the more prevalent) as a modular, bidirectional socket interface between an IP block and other interconnected blocks.

In this chapter we discuss several in-silicon bus-level debug environments based on on-chip instrumentation. These include bus-level monitors and bus-level trace, as they apply to fixed and socket-level traffic-monitoring approaches and as used for understanding the interaction and performance of the interconnect for real-time performance and bus-dependent processing operations and interactions.

6.1 On-Chip Buses

On-chip embedded bus management is typically more involved than for board-level buses. Board-level buses can tri-state the interface, which simplifies isolation; essentially, one only enables the chip-bus interface when it is needed. For on-chip busing, tri-state logic is much more difficult and expensive to implement. As a result, most on-chip buses are multiplexor based. That is, signals from bus masters are multiplexed at the master or slave sides of the bus interconnect to create a signal path between a bus master and slave.

N. Stollon, *On-Chip Instrumentation: Design and Debug for Systems on Chip*,
DOI 10.1007/978-1-4419-7563-8_6, © Springer Science+Business Media, LLC 2011

There are three fairly widely used standard on-chip buses; in order of widespread adoption, they are AMBA, OCP, and CoreConnect. Wishbone, an open IP bus interface, is also discussed for completeness. All these buses use similar interconnections and have a range of IP vendors supporting bus-compatible cores. There is also a variety of less standard and vendor-proprietary buses that continue to be used.

AMBA (advanced microcontroller bus architecture) is a family of bus architectures (which come in several varieties – AHB, APB, AXI) that is managed by (but not licensed as such) ARM Holdings PLC. The AMBA high-speed bus (AHB) is arguably the most widely used on-chip bus protocol, with multimaster arbitration, multilayer support, pipelining support, bursting support, and so on. APB is a simpler static bus architecture for peripheral systems. AXI (AMBA extended interface), the most recent AMBA variant, allows multiple outstanding transactions.

OCP (open-core protocol) is a bus architecture that is managed by the OCP-IP (international partnership). OCP defines a range of complex multicore and multichannel interfaces that address pipelining, multiple outstanding transactions, threads and tags, bursting support, and so on. OCP is based on a concept of socket-based interfaces that decouple the IP interfaces from the bus fabric to a large extent, allowing a large set of optional OCP interfaces in addition to a smaller configurable set of required signals. OCP also allows incorporation of user-defined interface signals to address application-specific requirements. The OCP debug signals discussed later are one example of a recently defined set of side band signals (which may be incorporated into future generations of the OCP standard).

CoreConnect was developed by IBM and is most widely seen in systems based on IBM PowerPC cores. It is also used by Xilinx as an internal bus architecture (in part because some high-end Xilinx parts have integrated PowerPC cores). CoreConnect defines a set of different buses – processor local bus (PLB), on-chip peripheral (OPB), and device control register (DCR) bus – for different applications. Each bus component of the CoreConnect architecture is optimized to achieve specific on-chip bus architecture goals. The PLB provides a high-bandwidth, low-latency connection between bus agents that are the main producers and consumers of the bus transaction traffic. The OPB provides a flexible connection path to peripherals and memory of various bus widths and transaction timing requirements while providing minimal performance impact to the PLB bus. The DCR bus is a mechanism for offloading system initialization and configuration, and related control related transaction traffic from the main system buses. The DMA controller and the interrupt controller cores use the DCR bus to access normal functional registers used during operation.

Wishbone is an open-community on-chip bus architecture. It is mostly seen in conjunction with freeware IP blocks. I am not aware of any silicon design that uses it, but it does come up in the literature and is popular with the IP freeware community.

6.2 Socket-Based SoC Design

Socket-based interconnect is a standards-oriented approach that focuses on adding value to the interface socket between the IP block and the bus fabric. Socket-based interconnect is an underlying principle in many OCP-based architectures, but it can also be applied to other bus architectures. Because many bus architectures allow addition and selection of various bus options that increase the functionality of the bus interconnect, using a socket-based interface simplifies addition, removal, or accommodation of the bus interface to the IP blocks, as well as the development of test suites to address verification and optimization of the design.

6.2.1 SoC Interconnect Complexities

Advanced buses allow a range of high-bandwidth implementations and define a number of features and capabilities in addition to baseline data transfer. These features include the extensions for special bus command modes, burst operations, and multiple data tags and threads that increase the number of traced signals. The flip side of working with advanced bus architectures is that they present an additional level of complexity when configuring and coordinating operation of large amounts of data. Analysis considerations include specifics of handshaking to a given interface and more global issues of how the on-chip bus subsystem is performing, such as understanding and optimizing bus transmission efficiency, latency, saturation, resource conflicts, and other operational considerations that can have a direct impact on the performance and operation of the processor components.

This visibility problem for the embedded SoC platform is more complex than can be addressed adequately by traditional on-chip test methods such as a traditional JTAG scan, for several reasons:

- Bus operations are multicycle, with signals in a bus cycle becoming active at different times, requiring sequential tracing, rather than as a single-cycle snapshot that a scan typically provides.
- Bus operation problems are interrelated with the operations of at least two communicating blocks (a processor and memory peripheral, for example). Traditional debug methods, such as halting part of a system for test, can introduce changes and new variables that interfere with the test scenario and process.
- If problems are intermittent or sparse, then trace operations must operate in a triggered mode, so information for a given range of bus cycles of interest is captured in real time.
- The problem is, to a large part, a multicore extension of embedded processor analysis, where run control and instruction execution and data trace are integral parts of processor support. For larger systems with multiple cores, the problem extends beyond processor execution to understanding system operation and communication.

All of this points to better understanding at the interconnect level being a critical layer of analysis. There are a variety of reasons why new generations of interconnects and analysis tools to support them are increasingly critical and important:

1. Heterogeneous multiprocessing ICs should efficiently handle complex data flow architectures with intercommunicating cores, with diverse requirements and features such as different data feeds, operating speeds, types of data endianness, diverse and dynamic levels of security, and quality of service (QoS).
2. Growing awareness that flexible and rapid integration of IP from multiple external sources is needed to reduce time to market, with concurrent requirements for integrating the test, hardware verification, and simulation environments.
3. There is a growing sophistication of the processors' data flow requirements, including the ability to handle multiprocessing and multithreading in efficient nonblocking manners. In particular, the multithreading features of leading-edge processors, from MIPS and others, benefit from both a processor and bus-level system analysis environment.
4. There is a growing appreciation for platform design approaches that efficiently address product upgrades, market segmentation, and product differentiation while maintaining common design infrastructure to keep design efforts manageable.
5. Supporting analysis IP provides a means of tying together pre-silicon and initial physical product verification by providing access and visibility to embedded operations. This analysis allows in-depth understanding of the design under different conditions.

Industry is addressing these issues with solutions that integrate both processor and bus trace for systems level debug (Fig. 6.1), which allow analysis of trade-offs and performance complex interconnect structures and socket-based IP integration.

At least three commercial companies offer interconnect and bus structure automation and IP tools, with several other SoC-centric interconnect approaches being used as proprietary customized solutions by SoC silicon vendors. Sonics (http://www.sonicsinc.com) offers the most mature commercially available solution, with its third-generation SMX and related SMART interconnect architectures. Alternative approaches include network on chip (NoC) interconnect architecture and a self timed (clockless) interconnect. All of these approaches rely, to varying levels, on a common concept of separation of the bus operations and core communications using socket-based interfaces.

Figures 6.2 and 6.3 show some of the features of a socket-based system design. Sockets communicate to initiator (master) and target (slave) interfaces, with functionality of the socket encompassing the necessary state machines, gating and multiplexing circuitry, and wiring to support desired data flow (including QoS, multithreaded nonblocking communication, security features, and dynamic power gating) operation. This allows for a more streamlined and compact bus fabric.

The socket consists of a set of agents that provide the signal and protocol management to address the specific interface needs of a core to the more general

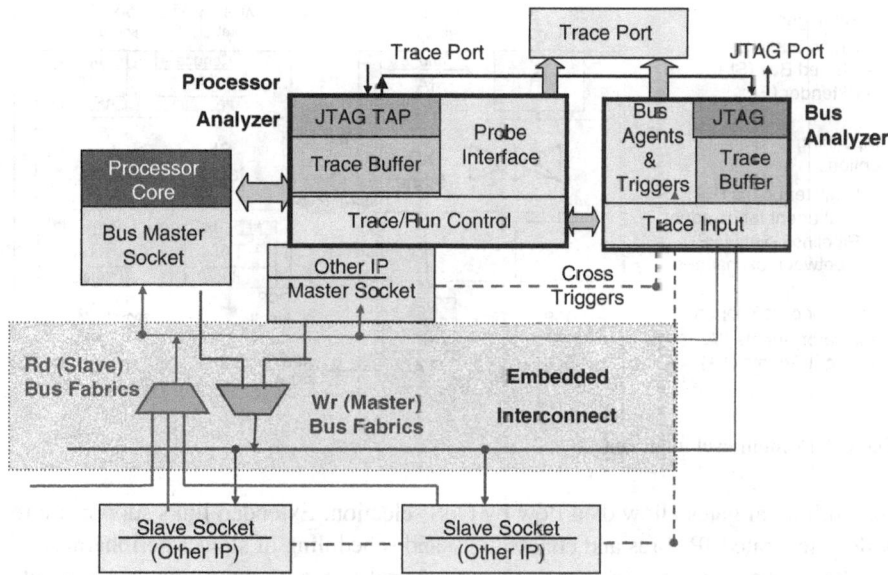

Fig. 6.1 SoC processor and bus trace

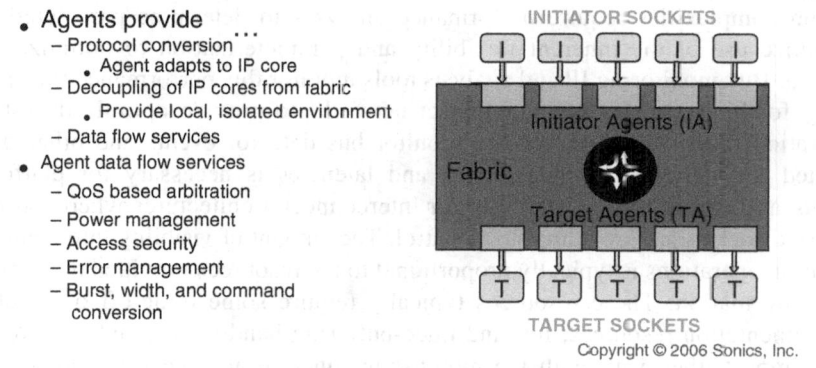

Fig. 6.2 Socket-based agents

resources of the interconnect fabric. Socket-based bridges can also define other interconnect linkages between OCP, AMBA AHB, and AXI and bridges for other arbitrary existing interconnect structures that can also be developed to simplify use of legacy hardware.

As a commercial example of such a complex interconnect fabric, Sonics' multi-service exchange (SMX^{tm}) can contain a distributed structure of three classes of interconnect structures, cross-bar exchanges, shared-link exchanges, and extended link exchanges, each with specific features and optimization requirements. Cross bars allow the fastest unimpeded connectivity, while shared links require less overhead

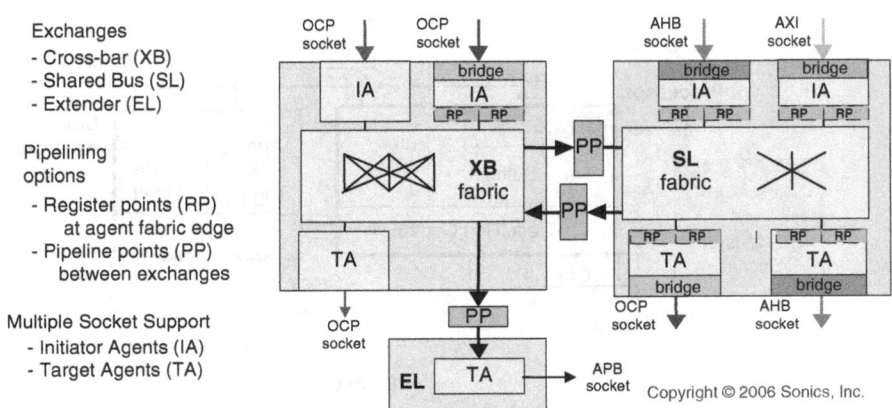

Fig. 6.3 A multilevel interconnect

and additional gates allow data flow by QoS selection. Extended links support more widely separated IP cores and connectivity and scheduling of slower peripherals.

These complex architectures support several types of interconnect segments that can be optimized for performance and require analysis information from the interconnect structure. Different types of interconnect segments have different integration and test requirements and communication features. Such communication complexities require performance analyses to determine parameters to optimize use of the inherent flexibility and parameterization to optimize the design. Bus-monitoring IP and analysis tools monitor this performance transparently for high-complexity interconnect networks to provide optimized system operation. Supporting the need to monitor bus data for events and other data related to intercore communications and latencies is necessary for platform debug and optimization, especially for interconnect architectures where parameterized sockets are providing flow control. The amount of visibility into communication operations is typically proportional to the resources provided to monitor key information. These resources typically require some trade-off of on-chip instrumentation resources, IO- and trace-buffering bandwidths, and the overall gate impact; they vary with the monitoring function and the size and performance of the interconnect structure. These trade-offs are discussed in the next section.

6.3 Bus-Level Integration

A widely used means of debugging bus systems is using monitoring mechanisms that are often included in creation of bus interfaces to detect incorrect addressing, illegal accesses, and timeout. Typically, the monitors are created as an option in automated bus-creation tools that create the bus fabrics and socket interfaces.

6.3.1   Bus Master Monitoring

Separate address-space monitoring is implemented for each bus master. If a bus master addresses an unused address space, the access is acknowledged with an error response and an interrupt is triggered. The incorrect access address and the associated access type (HBURST, HSIZE, HWRITE) and master ID are stored in a system control register. If more than one AHB master causes an access violation simultaneously (within a single bus clock cycle), only the violation of the highest-priority bus transaction is recorded.

6.3.2   Peripheral Bus Monitoring

The peripheral address space is monitored on the peripheral bus. If incorrect addressing is detected in the peripheral address space, access to the both master and slave sides is terminated. An interrupt is triggered and the incorrect access address is placed in a system control register.

6.3.3   Slave Monitoring

Bus slaves have limited responses to transactions; more typical debug concerns are with the timing of the peripheral response. There are three possible reasons for the timeout:

1. Actual timeout in the slave: If HREADY is still 0 after the maximum number of clock cycles, access to the master is terminated with an error response and the timeout interrupt is activated. The access to the slave continues. As long as the slave does not supply READY = 1b, all other accesses to the slave must be blocked with an error response. The interrupt is triggered only once. If the address phase of a non-IDLE access is pending in parallel to the extended data phase, this access is canceled and an IDLE address phase is output to the slave.
2. Too many retries in a row for the same access: Access to the master is terminated with an error response and the timeout interrupt is activated. Because there is no requirement that an access that has been rejected with retry has to be repeated, the next access of the master can be switched to the slave.
3. The SPLIT transaction is missing after a split response: Access to the master is terminated with an error response and the timeout interrupt is activated. The slave must continue to wait for signal HSPLIT = 1b. As long as the signal to the slave is missing, all other accesses to the slave must be blocked with an error response. According to the AHB specification, once the slave outputs HSPLIT = 1b, access must be repeated. However, because access is already terminated for the master, the data phase can no longer be handled correctly.

6.4   Internal and External Alternatives for Bus Trace

Debug ports simplify controllability and visibility by providing low-overhead access to internal signals. Debug interfaces can be categorized as either:

(a) Internal, in which most of the instrumentation functions are implemented on chip and the interface uses a low pin count interface, usually JTAG.
(b) External, in which the instrumentation functionality is shared between an on-chip component and an off-chip component, typically implemented in a probe that are connected by a (typically parallel) trace probe port.

Most JTAG-based instrumentation relies on on-chip memory to buffer between traced data and the available JTAG export bandwidth. The size of these buffers versus the amount of trace required is a trade-off because the amount of bus trace requires a large buffer. Buffers of modest size, however, are easily overloaded for a large amount of trace data generated in cases of multiple IP blocks or internal buses, placing limits on duration of trace that can be supported.

The underlying advantage of JTAG is that it is ubiquitous and is a default port implemented in most digital chips for test purposes. JTAG allows user-defined instructions to extend the JTAG instruction set for probe or trace modes, instructions for ICE, and to access internal JTAG-enabled registers. While in probe mode, the processor ICE can examine and modify the internal and external state of a system's registers, memory, and I/O space. In a trace mode, it can be used to serially export an arbitrarily large amount of information off chip. A rich infrastructure of tools environments and standardized debug schemes have been built on this foundation to provide JTAG debug of both embedded processors and other parts of an embedded system.

Adding an additional probe port provides IO bandwidth needed for more in-depth on-chip instrumentation approaches and is the primary focus of this chapter. Internal trace solutions and bus trace systems use an embedded trace solution, with an on-chip bus analyzer that is customized for bus analysis. On-chip RAM is a limiting resource, that can be spent on trace width of all or a portion of control, data, and address signals or trace depth, which can range from 64 to 64K trace cycles based on available on-chip RAM. Trace is controlled by user-defined combinatorial or sequential (state-based and counter-based) event triggering on trace or external trigger signals. This triggering can be used to disable trace until interesting events occur or to trigger on sparse or other irregular events of interest. Trace may include optional time stamping for multiinstrument synchronization or time marking for single-cycle or extended time traces. The same triggers can be used to drive debug-related actions such as cross-triggering between bus and processor or other IP operations.

A higher-performance alternative for bus trace is an off-chip mode that streams the bus trace to a high-bandwidth debug port. This allows a smaller RAM footprint by reducing or, in some cases (where trace bandwidth is less than or equal to the port bandwidth), eliminating the need for on-chip buffering. It also adds flexibility in what types and how much of the triggering and other supporting logic for bus trace are placed on chip rather than being inline or postprocessed by configurable logic in the probe. In general, off-chip trace attempts to minimize the amount of on-chip logic essential to triggering and filtering, leaving the trace intelligence (complex triggering, performance analysis, etc.) to be implemented in the probe and essentially using the using the trace port to funnel raw data to the probe as expediently as possible. The instrumentation features implemented in the probe allow a trace interface with a smaller simple trace logic and memory footprint and much deeper trace depth.

In many cases this trace port can be multiplexed with other bus functions. A trace port provides several advantages over JTAG trace and imposes other limitations. Although JTAG trace allows real-time trace constrained by the instrumentation and RAM speed, it requires some silicon area for the instrumentation and RAM (the size of which is largely proportional to instrumentation trigger features and trace depth, respectively) and has limited export bandwidth. The trace bandwidth capability for off-chip trace is typically limited by the number of IO pins dedicated to export of debug information at any given time, as well as the speed at which these signals can transmit the data.

Because the limitation on streaming trace requires more pins (for reasonable IO width) and may have maximum trace speed lower than operating speed (IO may have limited frequency performance compared to internal IP), it often makes sense to be selective on what bus data is being traced. Bus operations in particular are typically bursty in nature, and a bus may spend a significant amount of time in a quiescent state where no information is being transmitted. Simple filtering and buffering can significantly improve the usable bandwidth in an external trace solution.

In a more complex (multicore) instrumentation environment, external trace is limited by selection of critical data from different sources. The complexity of the funnel allows a range of performance trade-offs in external trace. By selectively choosing trace signals from different subsystems and instruments, an arbitration scheme that funnels the various trace information for export can increase effective trace bandwidth significantly. As shown in the following block diagram, the use of a trace port and JTAG is not exclusive. In many cases, JTAG remains important for control and configuration regardless of the trace mechanism.

6.5   Programmable Bus Performance Monitoring

As with most complex and heterogeneous systems, the ability to visualize and analyze performance characteristics is important to understand (and verify) system behavior, and subsequently fine-tune the system for optimal power and performance. Naturally, visualization and performance analysis capabilities are necessary during initial modeling and design verification and later, subsequent to chip tapeout in the lab, in order to observe the device running at-speed in the target system.

Such on-chip capability requires embedded instrumentation IP that allows embedded instruments to be used for a variety of functions, including performance monitoring, assertions, functional analysis, and debug – and even fault insertion and transaction stimulus. Although these functions have utility in hardware debug, they are more often used by software and systems engineers who can begin to leverage these on-chip resources to streamline a complex and time-consuming test and validation process.

A few additional requirements must be considered:

- Configurability: Given the configurability of modern buses, the embedded on-chip instruments must also be configurable
- Flexibility: The SoC will be composed of a variety of buses, interfaces, and IP blocks
- Easy insertion: Such a configurable and flexible system requires automated insertion

Ultimately, the solutions must be compliant to the OCP debug specification. However, even in the absence of IP cores or switch fabrics compliant to the OCP debug specification, practical performance monitoring implementations can be realized today.

The solution is a combination of programmable instruments, instrument programming, and analysis applications. The following is a description of an embedded-instrumentation solution delivering a comprehensive set of performance monitoring and analysis functions.

The overall objective is to provide the user with a spectrum of visualization and analysis methods, from coarse views on many interfaces (e.g. aggregate system throughput or worst-case latency) down through increasingly granular, targeted, and highly specific views at the socket level (e.g. discrete read/write transactions). This multilevel approach is consistent with many conventional analysis, diagnostic, and debugging methods; it seamlessly marries broad views of system behavior with "telescoping" views that are informed by the discoveries at each level of the visualization and analysis process.

This Socket level data can be accessed and extracted from multiple points in the design (Fig. 6.5), either through JTAG (Fig. 6.4) or trace ports (Fig. 6.6), using instruments applied to either buses or socket interfaces in the design (Fig. 6.7). Command/control of the instruments requires lower bandwidth and appropriate to a JTAG) interface. Perhaps the most important element is the programming and analysis application, which is often customized with a particular analysis view for different activities. In general tool views provide the user with a graphical interface and a set of high level commands to program and operate the on-chip instrumentation infrastructure (Fig. 6.8).

The basic suite of performance-monitoring functions includes the measurement of aggregate throughput per interface, master/slave throughput, instantaneous or average request/response latency, instantaneous or average event latency, and worst-case latency.

6.6 Bus Performance Monitoring

Performance monitoring functions can be realized with a distributed instrumentation scheme as shown in Fig. 6.9. There are three forms of instruments used: signal multiplexors, pattern match engines, and transaction engines. Each is a user-configurable

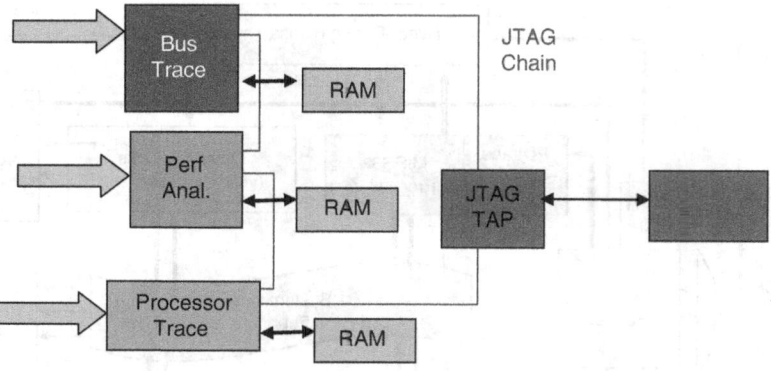

Fig. 6.4 JTAG-based internal trace

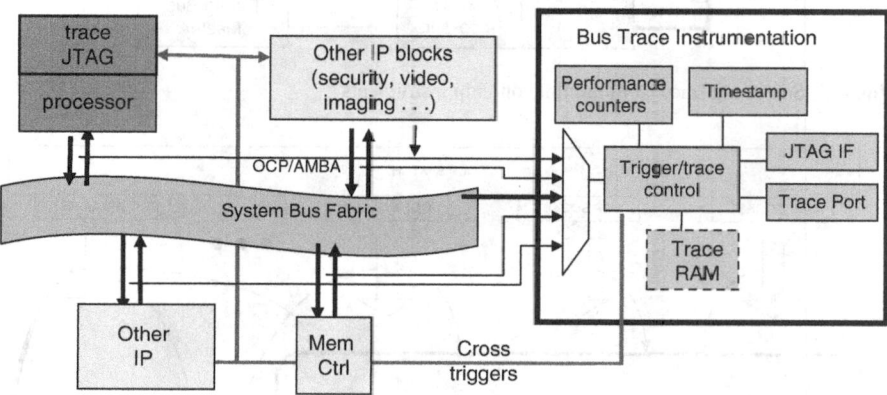

Fig. 6.5 Internal trace bus monitoring

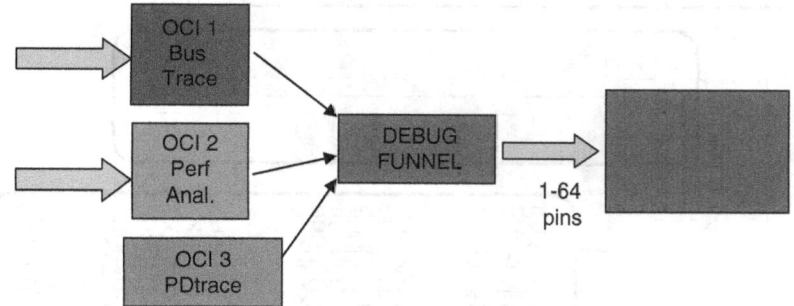

Fig. 6.6 External streaming trace

instrument (e.g. bus width, states, GPIO). All instruments are in-system programmable and run "at-speed" without disrupting the normal operation of the system. In this configuration, each multiplexor and pattern match engine operates autonomously so that multiple interfaces can be monitored concurrently.

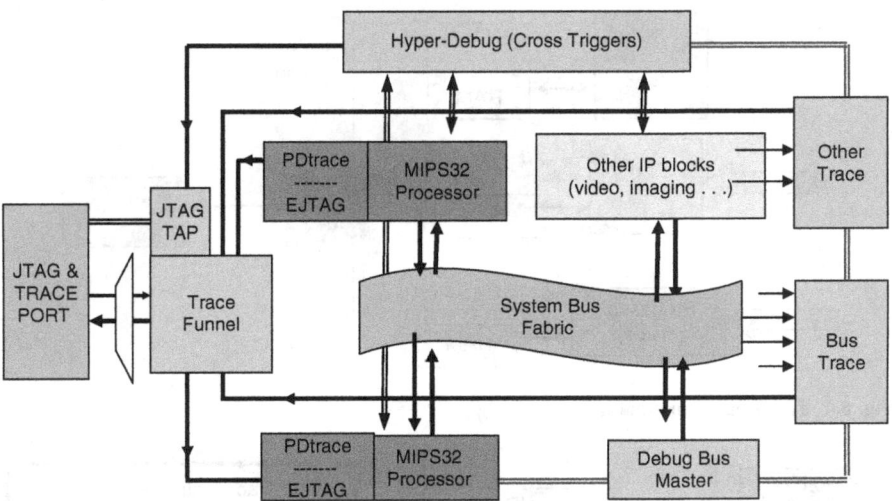

Fig. 6.7 Streaming trace from multiple on-chip instruments

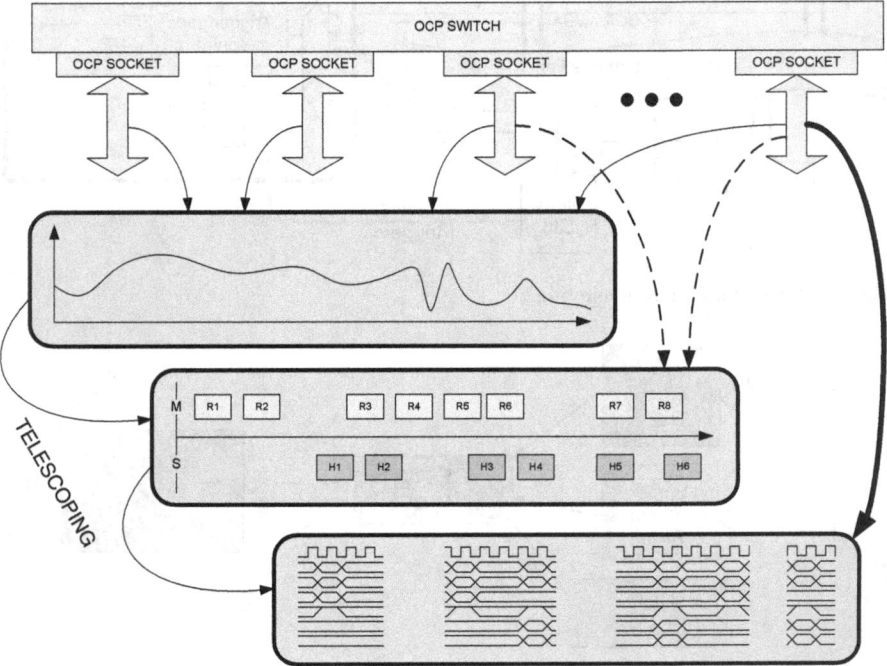

Fig. 6.8 Dynamic multilevel views

The multiplexors are used to reduce the number of signals presented to the pattern match engines and transaction engine. The advantage of multiplexing is that at runtime the user specifies which signals will be/should be monitored.

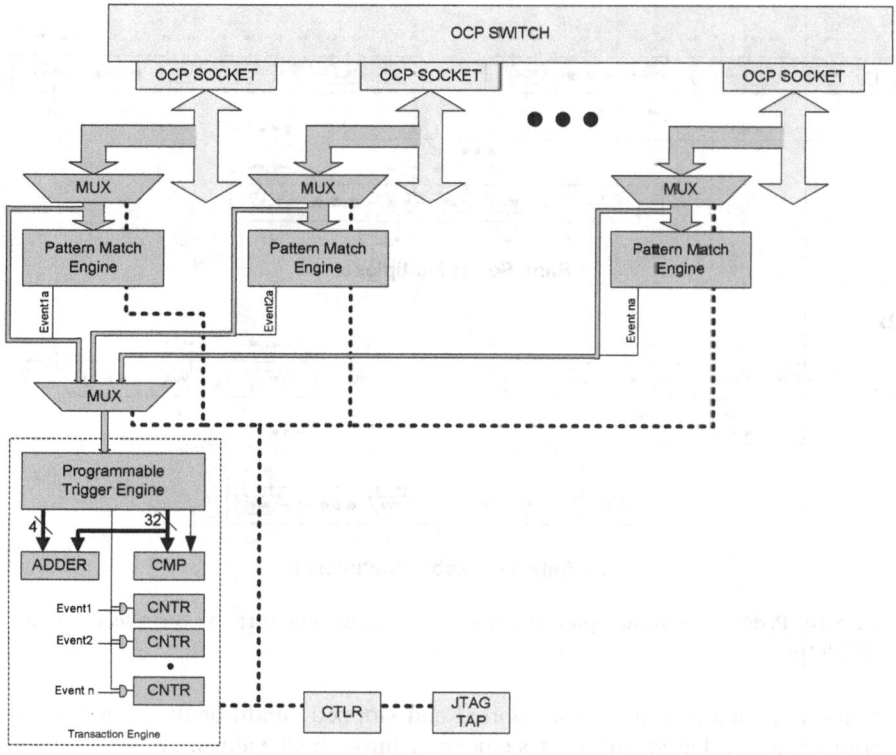

Fig. 6.9 Performance-monitoring instrumentation

The multiplexors can take on a variety of forms depending on the application, flexibility, and area overhead requirements. The bank select multiplexor is the smallest and least flexible (Fig. 6.10a). In this configuration, signal-bank-a or signal-bank-b is selected through a serially programmed register. The bit select crossbar multiplexor (Fig. 6.10b) offers additional flexibility by providing a serial programming register for each 2:1 multiplexor and an additional signal fan-out between multiplexor stages. This gives the user the most flexibility to select a greater combination of signals. In practice, the bank select multiplexor can be used in most performance-monitoring configurations, whereas the other configurations may be more appropriate if a variety of validation and debug functions are to be supported with the same instruments.

The basic pattern match engine is capable of detecting user-specified patterns on each interface. Whenever a specified pattern is detected, the event signal is asserted. The pattern values, mask values, and state machine configuration are specified at runtime within the programming and analysis application. The event signals are transferred through another set of multiplexors to the transaction engine.

The transaction engine provides a wide range of functions enabled through a rich set of resources that include a programmable state machine, comparators, counters, timers, and adders. The transaction engine can be programmed to count events, measure intervals between events, and measure frequency of

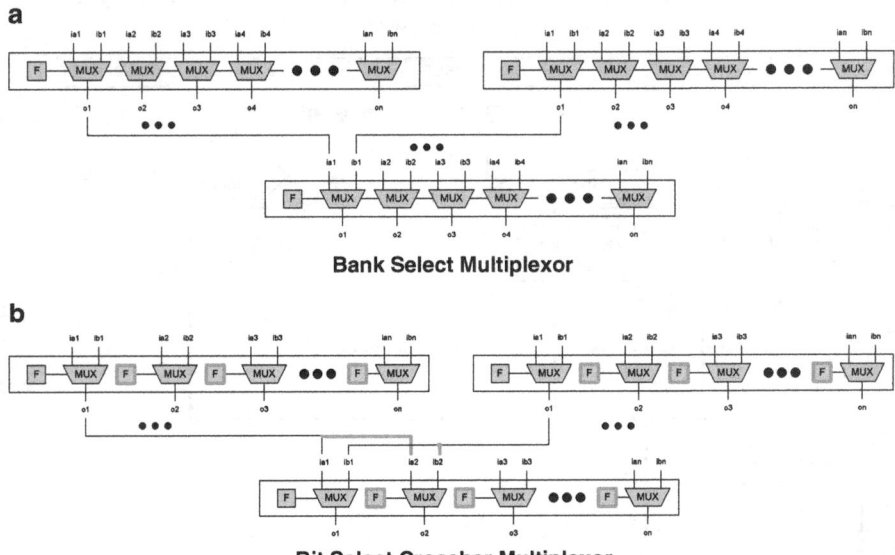

Fig. 6.10 Pattern match multiplexors (**a**) Bank select multiplexor; (**b**) Bit select crossbar multiplexor

events. All such actions can be started and stopped conditionally. Conditional actions may be based on event sequences, throughput values, counter values, latency values, or sideband events detected on signals mapped down to the trans-action engine through both multiplexor stages. In fact, even signals from other parts of the SoC can be used if they are available on the multiplexor inputs. The transaction engine may also be programmed with user-defined embedded memory for optional signal-tracing functions. All measurement values and calcu-lated results can be retrieved by the programming and analysis application for display and additional analysis.

6.7 On-Chip and Off-Chip Analysis

Given the obvious bandwidth limitations of the IEEE JTAG 1149.1 interface, large amounts of real-time data cannot be streamed off-chip. Although a high-speed trace port such as the Nexus interface (as described in chapter 11) can be used, many designs require a smaller interface. In the baseline configuration shown, we assume a system without a high-speed trace port. Such a configuration highlights a primary advantage of programmable instrumentation: the ability to perform on-chip analy-sis and reduce (but not eliminate) the amount of serial data transfer. Nevertheless, there are always situations that require off-chip analysis. This is especially true when on-chip data can be transferred into a variety of visualization and analysis

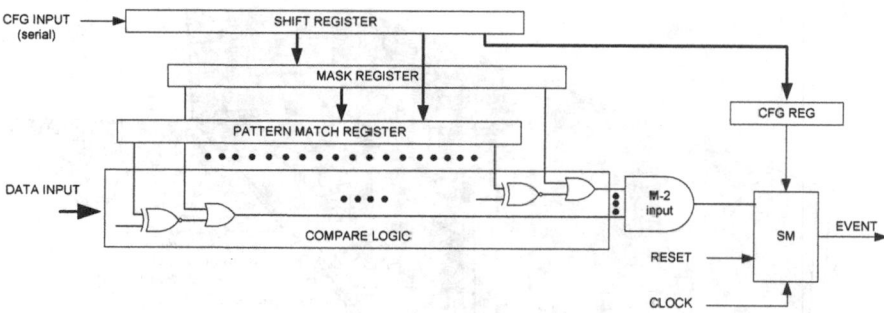

Fig. 6.11 Pattern match engine

tools. For example, the on-chip data can be retrieved and formatted into a trace file and is subsequently fed into analysis tools. Even if a subset of trace file fields is populated, transfer sequences and performance metrics such as throughput and latency can be analyzed (Fig. 6.11).

A bus monitor and analysis toolset allows performance, statistical, and transaction analysis of bus interfaces. Bus events are monitored and written to the trace format. A transaction re-builder reconstructs transactions from the trace files and builds up a data store. The transaction data can then be analyzed. User queries can be created, saved, and used in the analysis engine and the reporting tools. Performance statistics are calculated based on user queries. Reports are then automatically formatted and generated.

In complex and heterogeneous SoCs, the ability to visualize and analyze performance metrics is paramount to understanding and verifying system behavior and subsequently fine-tuning the system for optimal performance. That may entail not just looking at metrics over time but also having the capabilities to filter on various fine-grained aspects of the system and enable metrics to be viewed in a natural intuitive way. Figure 6.12 shows a 3D analysis chart of operation types versus cycles for a bus channel. The bottlenecks in the system can be viewed very concisely and the problem areas can be very quickly identified and understood.

In Fig. 6.12, the axes are time versus metrics versus channels. Zooming and filtering of the data can be done and metrics can be changed easily to allow infinite ways to view the data, depending on aspects that are trying to be understood or verified.

With the enormous amounts of data that must be captured in performance analysis, having a method for querying different types of data and metrics from the data store is useful in providing querying options that allow quick and easy access to the most critical areas of interest. Queries are not just limited to finding latencies, bandwidths, and other metrics above or below a certain threshold; they also have the capability of deep fine-grain analysis down to the transaction level. For example, particular bus transaction types on a monitored channel can be singled out and analyzed very quickly. Given the large timeframe of bus transactions, valid performance measurements should only take place at certain times, guided by certain events. For example, bandwidth and latency measurements are only really

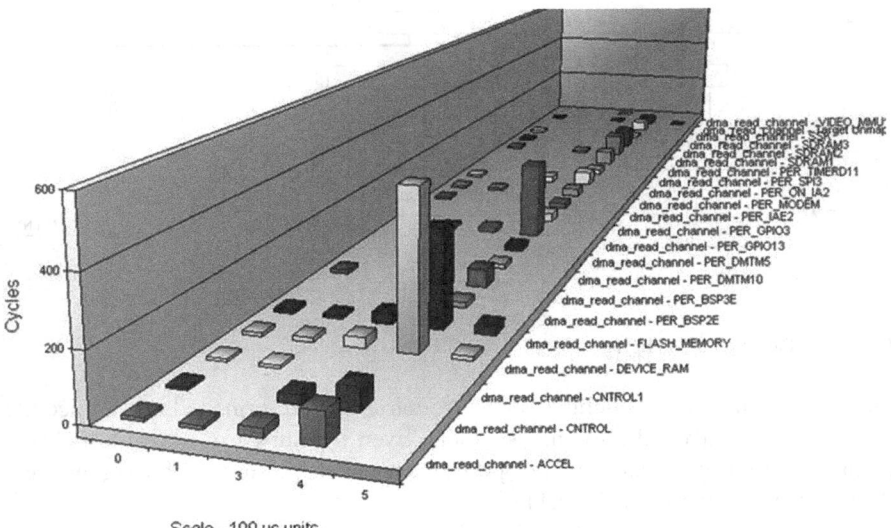

Fig. 6.12 A 3D bus analysis display

meaningful on a particular socket when the source traffic on that socket is correctly flowing. Event triggers are set up to enforce a valid measurement.

Events of interest vary widely and may include transactions occurring during a window; out-of-bounds events including DMA and SoC; and interrupt events indicating abnormal operations in timed events. A time event is simply a specified time defined by a timestamp or other counter for use as a trigger. Various triggers for event queries may include "Start After," "Measure From," and "Measure To." Each of these triggers can be one of the specified event classes.

Here again, the programmable nature of the instruments is beneficial. When it comes to extracting on-chip data, there is an obvious trade-off between temporal and spatial visibility. The amount of trace data to be captured is limited by the width and depth of the embedded trace memory. The user needs the means to make such trade-offs to maximize the utilization of the embedded memory. For example, if the user wants to see all activity on multiple bus interfaces, more transfer cycles may be captured if the data field is omitted. Likewise, the user may choose to reduce the number of transfer cycles captured by creating a capture filter based on a combination of address and command signals, or filter using signals associated with tag or thread extensions. Each of these techniques is accommodated by a simple expansion or reduction of the observation scope; the designer has ultimate control over the trade-off between temporal and spatial visibility. The programmable nature of the instruments allows these decisions to be made at runtime.

Although the configuration shown in Fig. 6.9 is suitable in many applications, more advanced configurations are possible. Consider, for example, scenarios that require multiple and specific *types* of transfers to be monitored simultaneously on

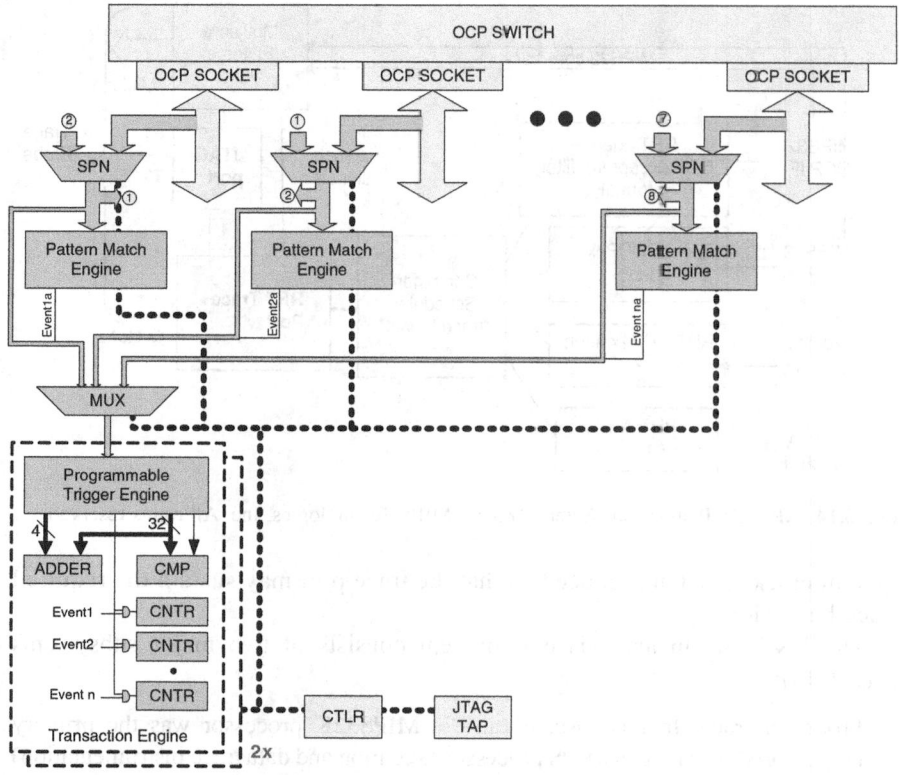

Fig. 6.13 Sharing pattern match engines

each bus interface. For such scenarios, two pattern match engines may be required. This can be accomplished in a variety of ways. A second pattern match engine can be dedicated to each interface, or adjacent pattern match engines can be shared between ports as shown in Fig. 6.13. Through the multiplexor configuration, the user has dynamic runtime control over the use of each pair of pattern match engines.

6.8 Request Response Trace Bus Analysis

Request response trace (RRT) was developed to analyze the complex data communication networks; both the data width and operating speed of the communication links can vary. Bus sockets vary from 32 to 128 bits, based on connection to specific cores. The bus fabric allows resolving the mixing of bus widths and speeds across different blocks, but efficiency and optimization of performance with regard to different data rates, clock rates, and other components of the system cores are not trivial. For the amount of trace required, core and

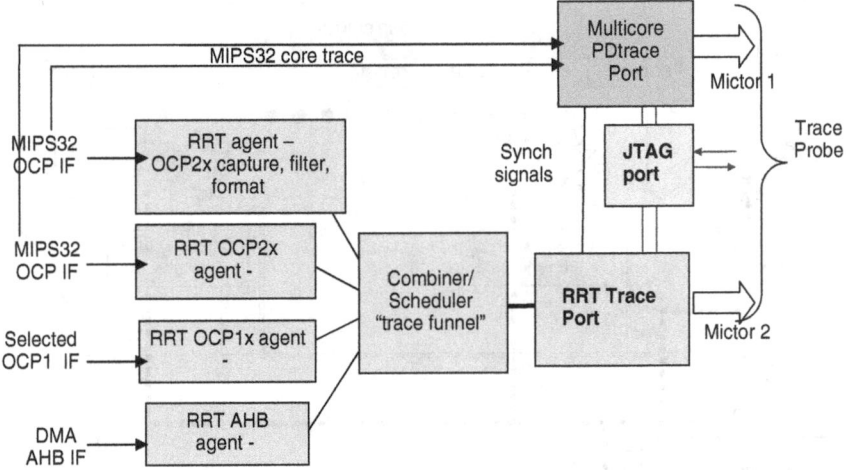

Fig. 6.14 RRT and PDtrace subsystem. *Source*: MIPS Technologies, Inc.

system clock speed are reduced so that the trace port may sustain the required trace bandwidth.

The RRT system analysis environment consists of two major subsystems (Fig. 6.14):

1. Processor trace: In this case, because a MIPS32K processor was the primary processor(s), PDtrace (a MIPS processor execution and data trace instrumentation) was implemented for each core. PDtrace interfaces support an aggregated processor trace port for both of the core trace outputs. Other processor selection would result in other trace systems being used for processor trace.
2. Bus-level request-response trace: This is a bus-level instrumentation system that ties into the bus fabric at the socket level and allows the trace of one single bus socket or all masters in the system simultaneously or a selection of masters. The RRT trace buffers each request-response output and includes a trace "funnel" to route the buffered outputs to the off-chip trace port.

RRT and PDtrace data are sent off-chip over a dedicated 16-channel trace port. Both PDtrace and RRT trace port interfaces are supported via a single probe, using two Mictor38 connector interfaces, each with its own independent clock source. The probe combines the trace inputs from the two sources and records them in a common memory buffer.

The probe includes a common JTAG connection and PDtrace trigger pins for trigger and trigger acknowledge. This trigger may also be used to put one or more cores in debug mode and to communicate with the processor and (in this example) the MIPS32 trace control and visualization tool, PDtrace. On-chip trigger output pins indicate to the probe the status of the processor core(s). The probe and on-chip logic have a common triggering methodology to allow the probe to enable and disable/stall RRT operations in conjunction with PDtrace operations. The triggering scheme also communicates stalling of the trace capture based on processor status. All applicable features of RRT and PDtrace, including the triggering, are configured via the JTAG port.

6.8.1 RRT Operations

The RRT provides for capture and collection (Fig. 6.16) of the following information. All capture is done on chip at the RRT agents and is exported via the RRT port:

(a) Recording of specifics of master-slave socket transactions and the number of clocks of delay between each request and response.
(b) Captured timing and latency of read cycles. Burst reads are reported on arrival of the first requested word or on the arrival of the last word of the burst.
(c) Transactions between one (selected) master and all slaves it transacts with, or several masters at the same time. These masters may include any of the following: two 34Kf cores, one active dedicated channel (selected as output of the crossbar), and one active channel of the DMA.

Trace collection allows overall capture for an extended (at least one video frame) processing period using the memory buffer in the probe (Fig. 6.15). Concatenation of multiple frames may be performed as a postprocessing stage on exported RRT trace files.

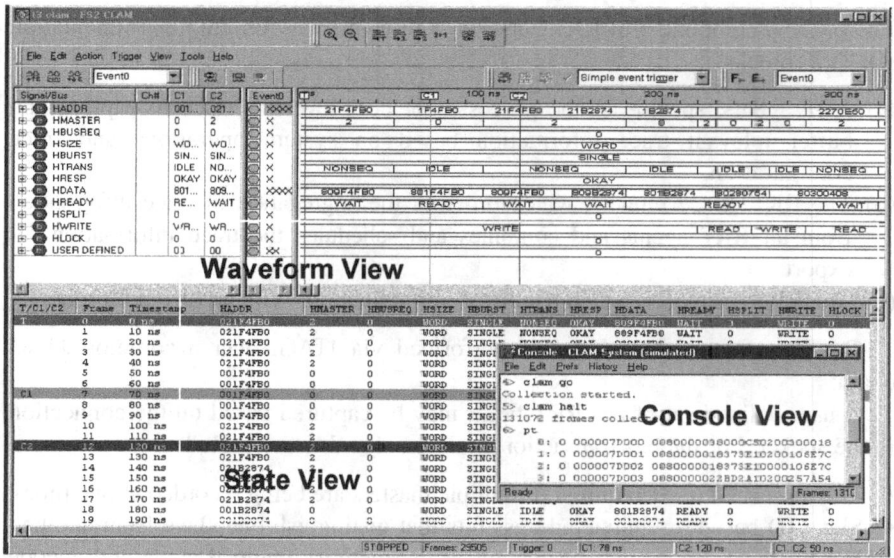

Fig. 6.15 RRT graphical bus trace. *Source*: MIPS Technologies, Inc. All rights reserved

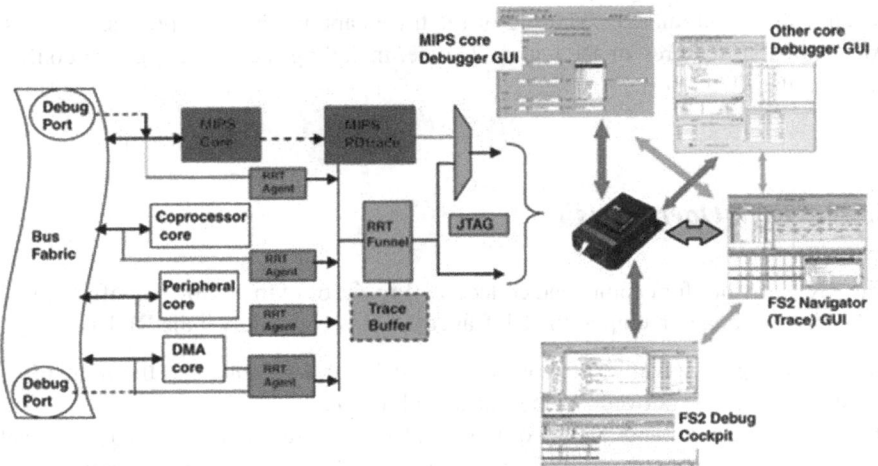

Fig. 6.16 Integrated bus and processor trace environments. *Source*: MIPS Technologies, Inc.

Post-trace software provides postprocessing and views of transactions and delay times over varying periods of time for both single and multiple cores. RRT data is correlated and used in conjunction with PDtrace data to provide a picture of system operation.

6.8.2 RRT Implementation

The on-chip component of RRT consists of three primary on-chip instrumentation (OCI) IP blocks:

(a) RRT agents, specific to the processor or core-level interface to capture and buffer relevant trace information based on system operations and trace configuration.
(b) The RRT "trace funnel", which provides the aggregation of trace information from all RRT agents and combines and schedules the trace information for export.
(c) The RRT trace port, which handles communications with an off-chip probe.

Configuration of each block is performed via JTAG, over a common JTAG chain:

A user-defined set of bus RRT fields may be captured based on the connection to the socket. For the bus transaction analysis, signals trace included:

- The master ID (only required if multiple masters are being recorded at one time).
- Slave ID based on unique address bits that distinguish one slave from another. Hardware in the agents can recognize the memory-mapped areas and encode them into the slave ID field.

- Protocol and traced bits that determine the alignment of a read response cycle to its "parent" request cycle.
- Request and/or response cycle type (or encoded in other fields).
- Cycle type – read versus write, single access versus burst.
- Buffer overflow indicator bit.
- Trace of upper address bits to determine code versus data memory-mapped regions. There are two defined modes: fast (partial) address field and full (complete address field) are user-selectable options.
- Trigger signal to allow on-chip subsystems to send a trigger signal to the probe.

To conserve trace bandwidth, the bus RRT records are further broken down into two modes: fast and full. Fast mode is limited to a single-cycle frame and includes socket-level control signals characterizing the bus transfer along with buffer overflow and/or trigger indicators. Full mode includes control signals as well as full address trace, based on a memory map of necessary upper addresses; typically transmitted over multiple trace clock cycles. The capture of this data via RRT allows the following to be performed during chip-level operation:

(a) Measurement of a processing loop such as frame time.
(b) Capturing available information for aligning socket measurements with core processor execution to correlate cause-effect of code execution to socket traffic based on coordinated recording of trace from both sources.
(c) Capturing available information on aligning socket measurements to correlate each hardware thread to the data transfers that each processor generates.
(d) Extraction of thread information extractable from socket address bits traced. Post-trace software can display per-thread socket transaction information providing valuable information to users on the density of transactions over time and the delays associated with those memory accesses, generated for each hardware thread.
(e) Postprocessing of the trace matches up requests and responses (using the socket protocol and possibly ID bits) and calculates the delay between them based on timestamp values stored along with trace to provide an accurate timeline of each request-response frame.

A RRT triggering system is implemented within the probe (off-chip) and includes event monitoring of all captured control and address signals to control start/stop and capture of trace information in the probe. This trigger may also be used to put one or more cores in debug mode and to communicate with the processor and PDtrace subsystems. On-chip trigger output pins indicate to the probe status of the processor cores.

The probe and on-chip logic have a common triggering communication to allow the probe to enable and disable/stall RRT operations in conjunction with PDtrace operations. The triggering scheme also communicates stalling of trace capture based on processor status.

RRT is supported by a set of control and display views and utilities to support analysis of RRT and PDtrace data. Additional visualization can be supported via export of trace to third-party tools. Control setup includes the setting of master trace priorities and selecting which masters are to be in the trace; trigger setup to analysis views for precise post-trigger positioning and reading trace and formatting data for additional analysis views.

Additional views include:

Raw State View for RRT: Basic acquisition is displayed as a state display that shows one line per trace frame with columns corresponding to the trace fields: transaction type (read/write, request or response), master name, slave name, transaction ID or outstanding request count, buffer overflow, and probe-generated trace timestamp values.

Aligned State View for RRT: Alignment *concatenates* two frames – a request cycle and its matching response cycle and a delta timestamp between the current and next transactions.

Graphical Display: Trace solutions are supported by a multiview (Navigator) GUI, which is customized for RRT data display as captured by the probe. The analyzer GUI allows complex triggering of capture and display of RRT information as waveform and state views. The GUI includes utilities for control of bus event monitoring and template-based triggering based on captured trace information.

Correlated View of RRT and PDtrace: Allows viewing of common PDtrace and RRT data captured at a common timestamp with a known or defined offset. It also allows RRT and PDtrace data to be locally correlated based on address values or common triggers, markers, and instruction (read/write/burst) types captured in both the PDtrace and RRT. Correlating socket traffic with instructions defines a processor to bus-level relationship, for example, by determining which thread caused a read or write operation on a given cycle to a peripheral socket.

Integration of the bus level and processor tools are integrated via a multicore API layer, which allows user transparent sharing at both the JTAG and trace port resources.

The instrumentation developed for the bus RRT system is designed to record request and response bus events at the socket interface and measurement of one processor (bus master socket) with expandability to allow concurrent viewing of key parameters of all masters simultaneously. This system analysis implementation allows capture of information about core load/store operations and their latency for the different socket masters, and exports them over dual trace ports to the probe, along with other trace and analysis data, in particular processor data interfaces that are used in providing complementary run control and trace analysis views of the processor operations.

Most bus-level analysis instrumentation and methodologies can be used across a range of architectures and bus interfaces; both the on-chip interconnect and analysis systems discussed can be applied to other processor cores or bus architectures under a similar generic scheme. The RRT tool chain has general application in several areas of SoC performance analysis and debug. RRT allows the real-time measurement of frame processing time in a SoC video processing system that used multiple cores and the interconnect bus IP. RRT was used to improve the utility of processor trace by capturing information for traces performed at the bus connection point and aligning with core processor execution, enabling correlation of the cause-effect of code execution to bus/socket traffic based on the time-based coordination of recorded traces from both sources.

RRT is also used in more general systems based on a multithreading processor. By tracing the bus interconnect to memory, RRT provides real-time bus latency information and metrics on if and when interconnect is stalled at the precise time of a requested load/store operation. However, this does not provide information on which hardware thread is running or on what part of the application code is running at that time. By correlating the trace information from the RRT bus socket(s) and the processor trace, it is possible to get a complete picture of how load/stores from each thread of execution are impacting overall system operation.

Leveraging this information one step further, optimization of the QoS system used to schedule threads in a multithreaded core operating within a SoC design can be achieved. For example, RRT can be used to extract thread information from socket address bits traced in a system. Post-trace software can then display per-thread bus/socket transaction data, providing valuable information to users on the density of transactions over time and the delays associated with those memory accesses, generated for each hardware thread. This detail of information is extremely useful for performance tuning the application of software threads of execution to the hardware threaded capability of a multithreaded processor, allowing system developers to optimize bus utilization and throughput in such a complex SoC design.

Chapter 7
Multiprocessor Debugging

Debugging becomes more complex when one introduces multiple processors. A peripheral that allows one to monitor bus activity between two processors, such as in an MCU+DSP device, can resolve shared memory contention issues. In traditional debug environments, one can only see what was written to a memory location, not which processor made the write. Bus-monitoring peripherals track the source of each memory access, providing the necessary information for the debugging environment to identify which processor made the write. This increased visibility adds complexity, so a debugger that can interleave the trace buffers between processors is needed.

In most designs, processors are integrated with several other subsystems that also may be included in systems analysis, such as trace operations. Logic blocks included in many designs include co-processors for specific applications, memory controllers, peripherals, and a host of other functions. Debug of these types of blocks can be supported by on-chip logic analyzers that allow triggering and trace of logic operations, which is often done in tandem with processor debug operations. One variant of logic analysis important for many systems is bus-level debug. Bus analysis typically takes one of two forms: signals of interest are traced at the bus interface (for example, an AMBA AHB port or OCP socket interface) or from within the selected debug points in the bus fabric.

Just as buses operate in conjunction with processors and other IPs, bus analysis must interface to other debug blocks. This is typically performed/accomplished/etc. using cross-trigger interfaces to the other debug blocks for low-latency triggering of the processor debug operations based on status in another core. Likewise, processor output signals can be used to allow triggering of other trace operations to start and stop based on processor operations. Cross-triggering resources, when combined with global timing control resources, such as timestamping of trace information allow synchronization and alignment of debug data from different sources being brought off chip. Concurrent trace permits a more systems-oriented focus on the debug process, by allowing simultaneous viewing of signals of subsystems operating in differing clock domains.

If a device supports bus monitoring, it will usually also support global breakpoints. With standard breakpoints, one processor can halt another processor only after a latency of several cycles. If the processors are out of synchronization with regard to interprocessor communication, this potentially aggravates debugging by requiring reset of both processors to a common resynch point. Global breakpoints halt both processors on the same cycle.

N. Stollon, *On-Chip Instrumentation: Design and Debug for Systems on Chip*, DOI 10.1007/978-1-4419-7563-8_7, © Springer Science+Business Media, LLC 2011

7.1   Cross-Triggering and Global Breakpoint Control

The cross-trigger block is distributed to all processor connections to a bus. If cross-trigger wiring is in the bus fabric, then only small wrappers (condition/action nodes) are required at each processor interface. Alternatively, a separate cross-trigger matrix may be implemented. The cross-trigger logic may be programmed from either a processor or a JTAG debugger. The underlying idea of the cross trigger is that any processor or significant on-chip logic block can generate an (edge or level) trigger output to other points within the chip, and receive cross trigger inputs from other cross trigger blocks on the chip, for subsequent processing or actions. The debugger or processor can configure specific trigger lines for each IP to send a condition signal (changing either polarity or delay) and enable or mask the trigger line from which it can receive a trigger/action operator (Fig. 7.1).

For example, we discuss cross-trigger subsystems that were included in the OCP-IP debug specification that allow event recognition from a combination of system-level and local (processor-specific) conditions and generate global or processor-specific actions based on the triggering of an event. In the first (HyperDebug), both conditional triggering and actions are dynamically controllable from system software. The HyperDebug block also provides a timestamping capability for trace and trigger synchronization of processor cores running in different clock domains.

The HyperDebug concept is simple: Accept a scalable number of status inputs from vendor cores and I/O pins, combine them in a user-configurable way, and send control outputs to a selectable set of vendor cores and I/O pins. HyperDebug is configured as a set of chains that connect condition nodes, which gather information about triggering events from different subsystems/cores of a design and about action nodes that distribute generated trigger outputs to different subsystems/cores in a design. The event-monitoring and triggering logic are handled in a HyperDebug control instrument (Fig. 7.2).

7.2   HyperDebug Distributed Cross-Triggering

HyperDebug connects to core signals through node agents called HyperDebug condition nodes (HDCNs). HDCNs for all cores that have status outputs such as run-state or trigger points connect together in a chain. Configuration bits at each

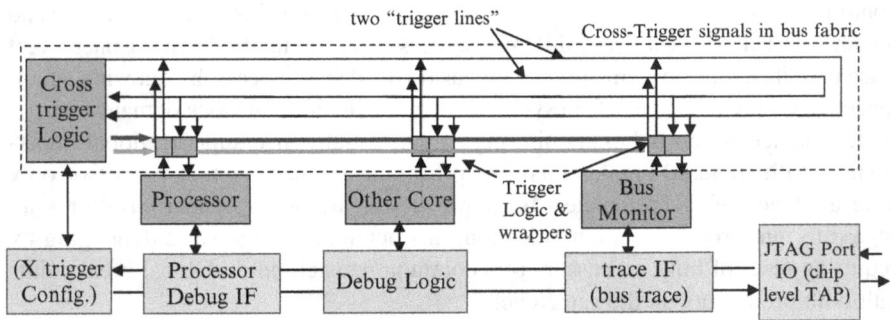

Fig. 7.1 Cross-trigger block diagram

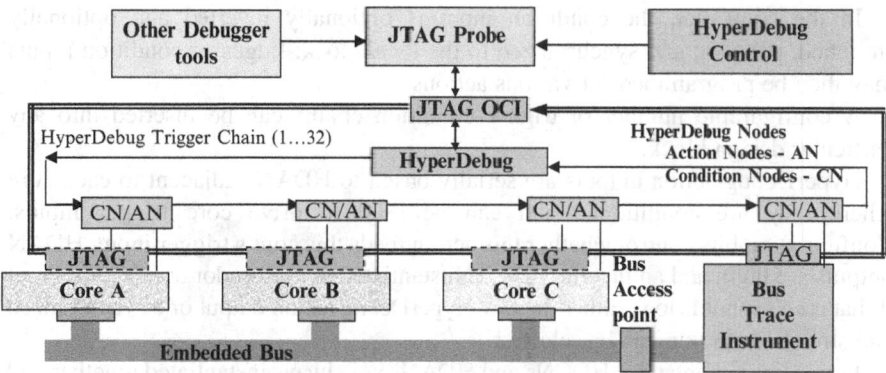

Fig. 7.2 HyperDebug cross-trigger block diagram

node optionally condition the logic of the core output and form a node output that is either a pass through from the node input or a combination of the node input with the node's status signal. In this way, a positive-logic AND or negative-logic OR combination of trigger outputs from any subset of cores in the chain is formed and feeds into the HyperDebug controller.

HyperDebug consists of three distributed types of components:

* HyperDebug OCI, which initiates the trigger condition and action operations and maintains the overall HyperDebug control and status.
* HyperDebug condition nodes (CNs), which modify the trigger conditions based on local conditions in the core, OCI, and other CNs connected to the core. Typically, a number of CN blocks are implemented related to trigger conditions monitored in a given core.
* HyperDebug action nodes (AN), which initiate logical actions such as setting registers in the core or OCI. AN operations are local to specific codes or may be global to all cores in the SoC (halting or resetting the core is one example). Typically, the number of the AN block is related to the number of actions that would be required for the debug logic to control core operations.

The OCI accepts a configurable number of condition inputs and generates a configurable number of action outputs. Condition inputs may come from a chain of CNs or from external pins fed from a JTAG probe. Action outputs may go to a chain of ANs or to external pins leading to the JTAG probe. The number of conditions need not necessarily match the number of actions.

Condition inputs are synchronized and stretched to match the clock period of the HyperDebug OCI. HyperDebug trigger conditions are an AND combination of one or more of the following:

* Condition input from a CN chain.
* Condition input from an external pin.
* An event counter matches a pre-programmed value.
* HyperDebug sequencer state.

In the controller, the condition input is optionally inverted and optionally stretched, delayed, and synchronized to the local clock. Edges of condition inputs may then be programmed for various actions.

A configurable number of trigger condition chains can be inserted into any particular design block.

HyperDebug action outputs are serially bused to HDANs adjacent to each core where they are conditioned and can be used to drive core trigger inputs. Configuration bits control whether this action feeds this core's trigger input. HDAN outputs are high, and an inverter may be instantiated at the vendor core action input if that core's input is low. Either the raw HyperDebug action output or a synchronized and stretched version can be selected.

In this implementation, HDCNs and HDANs are always instantiated together and the combination of an HDCN and HDAN at a particular core is called a HyperDebug Node (HDN). An HDN connects to one condition and one action chain. The logic for nodes can be very simple or more complex depending on the level of triggering complexity required, including state machines for sequential triggers. A simpler combinatorial implementation with programmable delay is often sufficient.

A configuration clock (the TCK input and configuration enable signal) is used to initialize the configuration registers in the HDN. The configuration chain is similar to JTAG in that at a rising TCK edge, the hd_condition_out output is latched into the first register in the chain (like TDI), while the bits in the HDN configuration register are shifted one bit forward. Output from each HDN changes on the falling edge of the configuration clock (like TDO) so that routing delay and clock skew between HDNs is not an issue. When the chain is not in configuration mode, hd_condition_out supplies the logic 1 feeding into the first HDCN as illustrated by the Figures 7.1 and 7.2.

If a core or the chip I/O has more than one potential trigger status output or action input, more than one HDN may be instantiated at that core.

7.2.1   HyperDebug Controller

The OCI accepts a configurable number of condition inputs, and generates an action output for each. Condition inputs can come internally from a chain of HDNs associated with cores or externally form inputs from external instruments or other logic. The Action outputs propagate through the HDN chain to cores or external pins.

Condition inputs are synchronized and stretched to match the clock period of the HyperDebug OCI. Any convenient clock can be used to drive HyperDebug. HyperDebug trigger conditions are either:

- Condition input from the HDN chain AND the HyperDebug sequencer state.
- A global event counter matches its preprogrammed compare value AND the HyperDebug sequencer state is one of a specified list.

When a condition is indicated, the HyperDebug controller may be programmed to perform actions:

- Assert, negate, or pulse the action output to the HDN chain.
- Start, stop, increment, or clear the global 32-bit event counter.
- Change to another HyperDebug sequencer state.

The condition input to the HyperDebug controller is conditioned according to three parameters set up by the user:

- The user can select whether to invert the condition. The optional inversion would be used when the condition is active-low, such as when the bus is used as a logical OR of several cores.
- A synchronizer is used when the sequencer clock domain is different from that of the cores or if the routing delay of the condition bus is significant compared to the clock period. For small systems, the synchronizer may not be needed.
- An edge detector would be used after the synchronizer to change the duration of a condition to one clock. For example, this might be used if a core is set up to assert its condition output when a certain trigger point occurs and the user would like the instruments to break after a certain number of these trigger points have occurred. Each trigger point asserts the condition bus for one core clock, and the condition bus is then resynchronized to the clock. The edge detector guarantees that the counter increments once for each trigger point even if the core's condition output lasts longer than one clock (Fig. 7.3).

7.2.2 Typical HyperDebug Implementation

When a condition is indicated, the HyperDebug controller instrumentation may be programmed to perform one or more actions:

- Assert, negate, or pulse one or more action outputs to an AN chain.
- Assert, negate, or pulse one or more action outputs to an external pin.

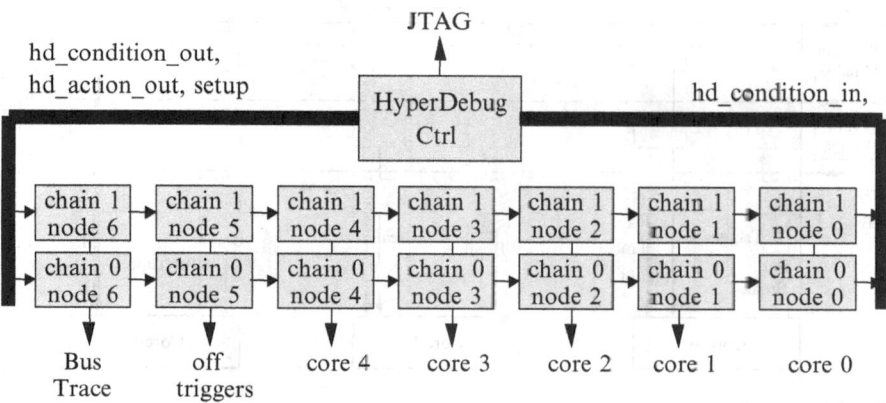

Fig. 7.3 A HyperDebug configuration

- Start, stop, increment, or clear operations on a 32-bit event counter.
- Change to another HyperDebug sequencer state.

The HyperDebug block also sources a reference clock signal for timestamping of data at each OCI block. Given that cores in a system may operate over a range of frequencies, including asynchronously to each other, a master timestamp provides a means of synchronizing the time of a core operation in relation to other cores in the system.

A typical implementation of HyperDebug with two HDN chains can support the following capabilities:

- Sequential, multicore triggering, such as event A followed by B or trigger 1 ms after an event.
- Periodic trigger signal to insert synchronization messages in each core's trace buffer.
- Assert a logical break signal to all cores under debug when any of the cores hits a breakpoint.
- Insert a trace message in each core's trace buffer when a particular core reaches a trigger point.

With multiple chains, conditions for cores corresponding to different problem domains (concurrent breaks or interupts, power control, trace, etc.) can be assigned different chain connections and operate concurrently (Fig. 7.4).

Trigger-out and trigger-in routing can be handled as sideband signals by the bus interconnect. The cross-triggering programming can be handled at subsystem level via the Hyperdebug control block trigger event, which can also be routed to the trace components via action/condition nodes. Trigger events can generate either a debug request or an interrupt request. These differ for different cores.

The cross-triggering can support external triggers. The trigger pulse width must be compatible with device IO performance. Triggers can connect to IO. Level or

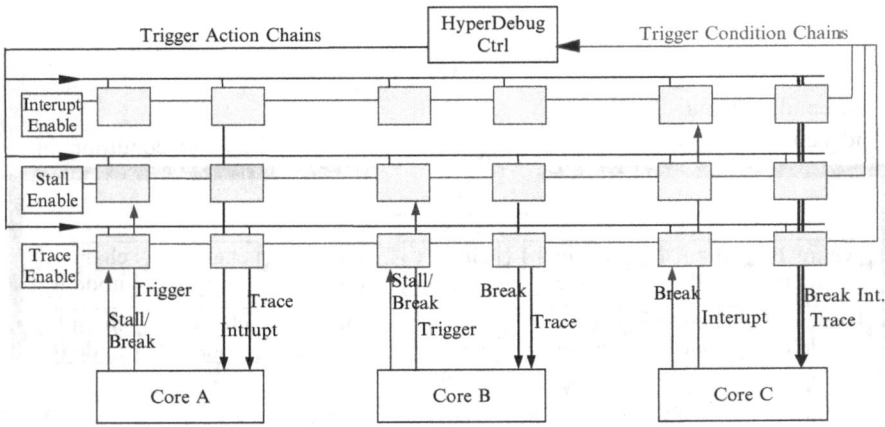

Fig. 7.4 HyperDebug configuration

pulse triggers are supported. A subsystem in power down or where debug has not been enabled does not contribute to cross-triggering.

The cross-triggering supports independent clock domains for a trigger-out master and a trigger-in target. The cross-triggering must be operational for any platform subsystem's frequency operating point, assuming a simple action/condition node configuration.

7.3 Multicore Synchronization Triggering and Global Actions

The amount of information in a multicore SoC is large enough that global event recognition is often needed to identify and isolate events occurring throughout the system. Event recognition is widely used in conjunction with trace to capture information on events and operations in the SoC. Trace data values are monitored and compared to event sequences to provide real-time triggers in instrument block(s). These triggers in turn can be used to control event actions such as breakpoints and trace collection. Multicore debug instrument event recognizers can simultaneously look for bus address, data, and/or control values and be programmed to trigger on specific values or sequences such as address regions and data read or write cycle types (Fig. 7.5).

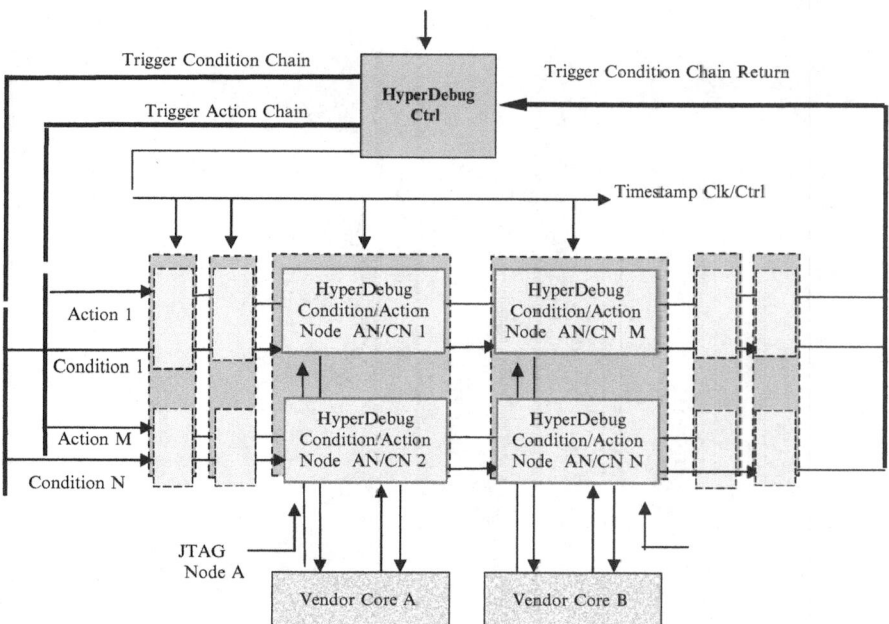

Fig. 7.5 HyperDebug system integration

Chapter 8
IEEE 1149.7: cJTAG/aJTAG

IEEE 1149.7 (Standard for Reduced-Pin and Enhanced-Functionality Test Access Port and Boundary Scan Architecture) is a superset of the 1149.1 JTAG interface, which, as previously discussed, has been in use since 1990. The IEEE 1149.7 standard (also known in the past as cJTAG or Compact JTAG, and later as aJTAG or Advanced JTAG due to copyright issues with the name CJTAG) was developed to address some of the known limitations of 1149.1 performance and extend its capabilities by creating a complementary standard that addresses the changes in the integrated circuit technology and topology. Originally defined as part the MIPI Test and Bdebug Working Group, 1149.7 defines a hierarchy of new, but JTAG compatible capabilities. With increasing levels of complexity, it replaces the JTAG TAP FSM with alternate TAP.7 architectures to implement additional functionality and maximize debug performance. Like other JTAG extension, a key concern is to maintain backward compatibility with IEEE 1149.1 infrastructure, semiconductor IP, software IP, and existing debug and test tools. Additional functionality and improved debug capabilities include

1. Provide mechanisms for TAP power management.
2. Provide modes that provide test and debug capability with fewer pins.
3. Provide background instrumentation capability using the same pins.
4. Preserve a gateway for debugging semiconductor errors/defects.
5. Provide a framework to improve debug use performance and allow other debug pin protocols to gain access to the pins.

The 1149.7 architecture maintains the underlying JTAG-compliant control mechanism while providing several extensions to JTAG. Instead of a radical departure from the existing standard, 1149.7 represents the natural evolution of the standard.

The 1149.7 operating mode is JTAG-compliant from power-up. Conventional JTAG control sequences are used to switch between the JTAG-compliant standard mode and the 1149.7 advanced mode. A 1149.7 aware debug and test software environment (DTS) must therefore be able to determine whether it is communicating with existing JTAG components (legacy), a mix of legacy and 1149.7 components, or a system built with only 1149.7 components. It can also determine whether the components are configured in a star or series configuration.

N. Stollon, *On-Chip Instrumentation: Design and Debug for Systems on Chip*,
DOI 10.1007/978-1-4419-7563-8_8, © Springer Science+Business Media, LLC 2011

The 1149.7 architecture supports the following features:

- 2-pin operation (compared to the 4 or 5 pins of standard JTAG or the 5 or 6 pins if the JTAG return test clock [RTCK] is included). The latter interface is referred to as modified IEEE 1149.1.
- Target operating frequencies (TCK) from DC to 100 MHz.
- Compatible with all hardware/software that uses the JTAG standard.
- Provides debug access that is independent of PROCESSOR and debug technology.
- Supports multidevice communications ports with up to 16 devices per port.
- Creates data transport channels superimposed on JTAG stable states such as idle and the two pause states. The stays in these states may be used to move background or custom instrumentation data using 1149.7 or private protocols (BDX/CDX).
- Power domain awareness at target and board levels.
- Comprehends synchronized operations across multiple debug ports.
- Tolerates slow system response such as power save modes or component clock limitations.
- Includes failsafe and robustness features.

The 1149.7 architecture builds on existing technology and legacy hardware/ software. This evolutionary approach maintains the value of the vast majority of IP created since the JTAG standard's inception.

There is a facility within 1149.7 that allows hot connect to the target system without system disturbance. This facility, called firewall, "disconnects" the JTAG devices from the TS 1149.7 adapter by gating off the adapter's TCK output to the connected JTAG devices. Debug software can use a standard JTAG sequence to disable this firewall.

Another capability, called super bypass, may be used in the JTAG mode of an 1149.7-enabled chip. Super bypass provides a one-bit bypass between TDI and TDO for both instruction and data scans, thus reducing the scan path length in a system.

Another 1149.7 capability, the BDX/CDX (background data transport/custom data transport) mode, can be used to transfer data between the DTS and target system. BDX allows transfers during idle control periods. CDX transfers control to the TAP on-chip logic to control and initiate transfers with off chip elements. Transfers occur when stable TAP states are reached (e.g. idle, pause-IR/DR, or shift-DR/IR). These transfers support, for example, user I/O, outputting instrumentation trace information, and custom protocols such as those for non-JTAG debug technologies.

8.1   Test and Debug Views of 1149.7

Because JTAG-related TAPs have become the most common debug port, improvements in providing access to chip facilities that support application debug and system integration is a higher priority than it was when. 1149.1 was developed and on-chip

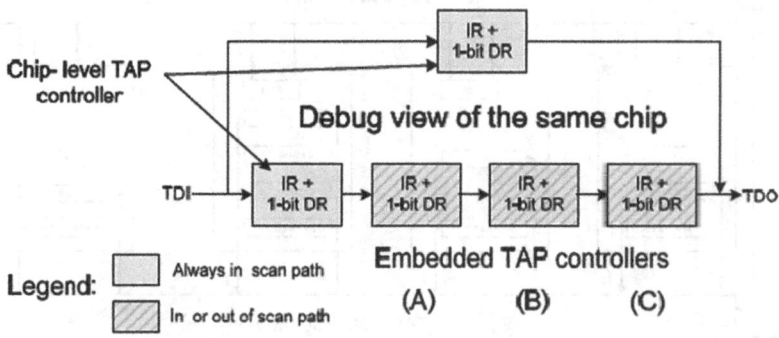

Fig. 8.1 Test and debug view of a TAP. *Source*: IEEE 1149.7

debug was a low-priority factor. With on-chip debug environments for different on-chip processors and other logic required in complex SoC, the JTAG TAP must be able to provide application debug support and access to multiple on-chip TAP controllers and embedded TAPs for special-purpose debug logic. This increases the complexity of applying JTAG to debug environments, compared to the relatively simpler and more consistent test requirement of scan operations.

1149.7 defines a new controller architecture (TAP.7) that includes the 1149.1 controller for compatibility modes, but depending on more advanced modes of operation, expands the functionality and relationships of JTAG states and modifies the requirements of JTAG-required registers to allow new functions.

Test applications do not require visibility to the on-chip system components that are of interest to debug, and debug does not typically require visibility to the interconnection between the chips needed for a board manufacturing test. This means that test and debug desire two different views of a system of interest, as shown in Fig. 8.1. The views are complementary, because testing is generally completed before application debug begins using different tools and design for test software.

1149.7 addresses this need for both test and debug views in its definition of a TAP.7.controller. The test view requires compliance with the IEEE 1149.1 standard. The debug view provides access to multiple TAPs within a chip, as shown in Fig. 8.2. These are very different views of the same logic.

The test view is given initial priority over the debug view. A testlogic reset TAP control state creates the test view. This view remains until an action is initiated via the chip-level TAP control to change the controller view. When the debug view is used, the chip creates visible subsystems with debug components that may be accessed. In most cases, the TAP control associated with an on-chip component controls the component but does not have boundary scan associated with it.

The debug-related and auxiliary TAPs within a chip may be included and excluded from the scan chain depending on their availability (some may be powered down or otherwise inaccessible). The inclusion and exclusion of TAP

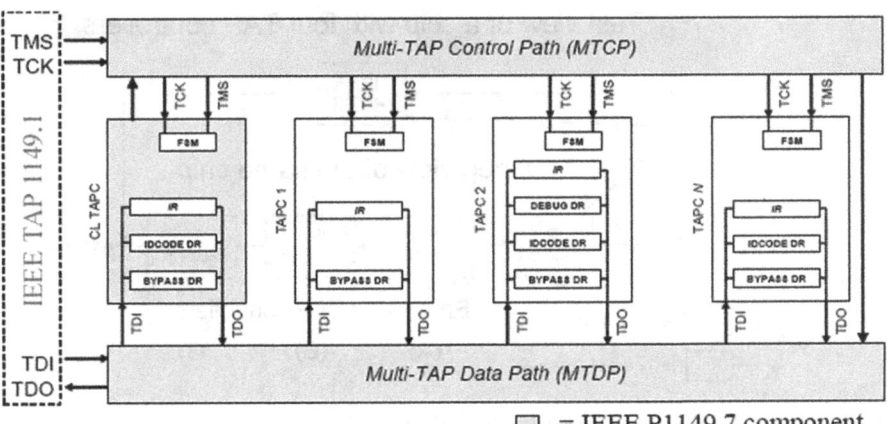

Fig. 8.2 Multi-TAP control and data paths. *Source*: IEEE 1149.7

controls from a chip scan path occur upon entry into the run test/idle TAPC state. This allows the synchronization of actions such as running or halting processors attached to different TAPs with a chip, similarly to how JTAG would address different chips within a system.

8.2 Key T0–T5 Class Functions

The key IEEE 1149.7 features are addressed in six class functions (designated T0–T5) that are defined through 1149.7. Classes T0 through T3 extend IEEE 1149.1 and enable new operations. Classes T4 and T5 are focused on advanced two-pin operation.

Class T0 ensures compliance with current test infrastructure by configuring IEEE 1149.7 devices to make them act compatibly with IEEE 1149.1, as shown in Fig. 8.3. 1149.7 defines supporting features, many of which are optional or undefined in 1149.1, including the use of N-bit IR, 1-bit DR for bypass instruction, mandatory 32-bit IDCODE, and mandatory instructions behaving as specified in IEEE 1149.1. After a testlogic reset is initiated, all multi-TAP devices must conform to the mandatory IEEE 1149.1 instruction behavior and implement a 1-bit DR scan for the bypass instruction.

Class T1 instantiates a control system for the IEEE 1149.7 standard that is transparent to IEEE 1149.1 devices, providing a foundation for the advanced functionality implemented in classes T1 through T5 without changing the IEEE 1149.1 state machine. In addition to creating a control system based on a TAP7 controller and extended protocol unit (EPU), shown in Fig. 8.7 and the commands and registers associated with direct addressability for classes 1 to 3, this class addresses the needs of power-sensitive devices with four power-down modes.

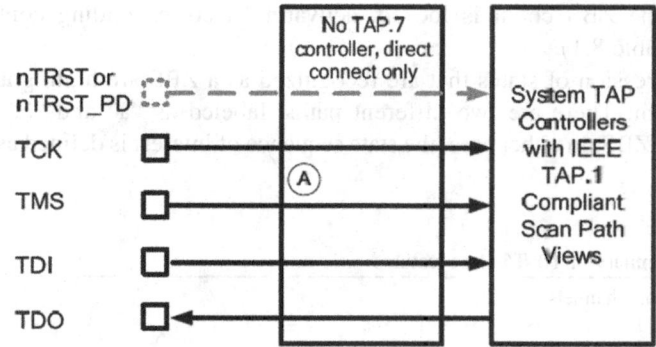

Fig. 8.3 Typical T0 class TAP.7. *Source*: IEEE 1149.7

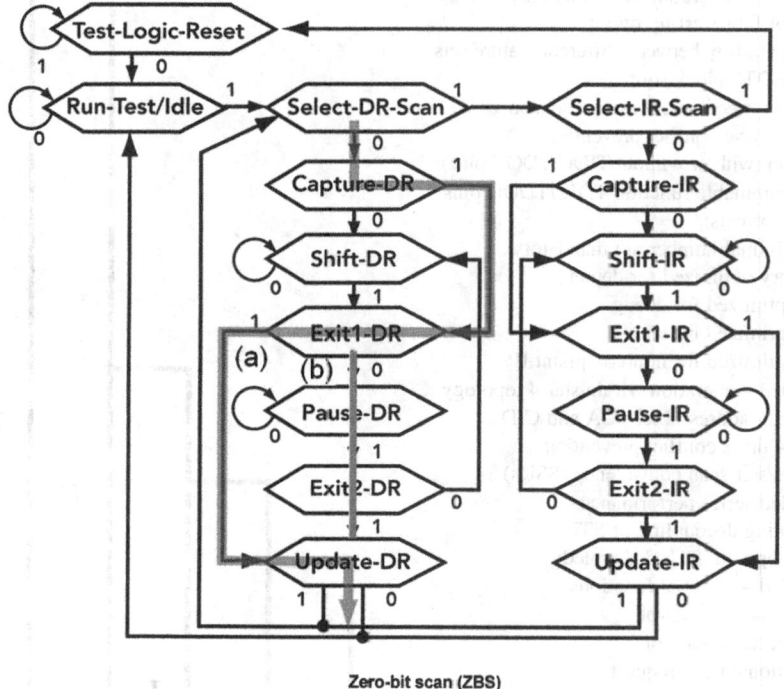

Zero-bit scan (ZBS)

Fig. 8.4 The 1149.7 modified JTAG FSM with ZBS paths highlighted. *Source*: IEEE 1149.7

The key innovation is the combination of the IEEE 1149.1-compatible TAP state sequences and shift state watching, which creates an IEEE 1149.7 control system that uses the bypass or IDCODE instructions plus a series of IEEE 1149.1-compliant sequences called zero-bit DR scans (ZBS), shown in Fig. 8.4. Beginning at zero, the ZBS count is incremented with each consecutive occurrence of a ZBS without encountering a shift-DR TAP controller (TAPC) state. When a DR scan containing a shift-DR occurs and the ZBS count is greater than zero, the ZBS count is locked, activating a corresponding control level (shown in Table 8.1).

The progression of states that are recognized as a ZBS are highlighted in the FSM diagram. There are two different paths, labeled as "a" and "b," that can implement a ZBS. In either case, the state sequence of interest is defined as follows:

Table 8.1 Summary of T0–T5 class features

Advanced – data channels
  Data channel 1
  Data channel 0
  No data channels, don't go office
  BDX/CDX transfers
Advanced – operation within star-2 topology
  One of four start-up options
  Compatibility between different feature sets
  TS or DTS clock source
  Test reset equivalent escape sequence
  Star-2 drive conflict prevention
  2/4 pin (with or without TIDC/TDOC pins)
  Programmable function TDIC/TDOC pins
  Scan formats:
    –Minimal number are mandatory
    –Very optimized for debug
    –Optimized for debug
    –Optimized for test
    –Optimized for non-compliant IP
Extended – operation within star-4 topology
  Directly addressable, TCA and CIDs
  Star-4 drive conflict prevention
  Series/star scan equivalence (SSDs)
Extended series performance
  Coupling/decoupling of STL
  Start-up with STL decoupled
Extended – optional functions
  TAP.7 power control
  Test reset generation
  Functional reset request
Extended – control levels
  Control level two – Cmds. And Regs.
  Control level three reserved
  Control level four/five scan paths
  Control level six and seven DTS use
1149.1 compliance
  IEEE 1149.1 compliance at start-up
  Multiple embedded TAPs
  Coupling/decoupling of embedded
    TAPs: Inclusion and exclusion of
    DR scan paths

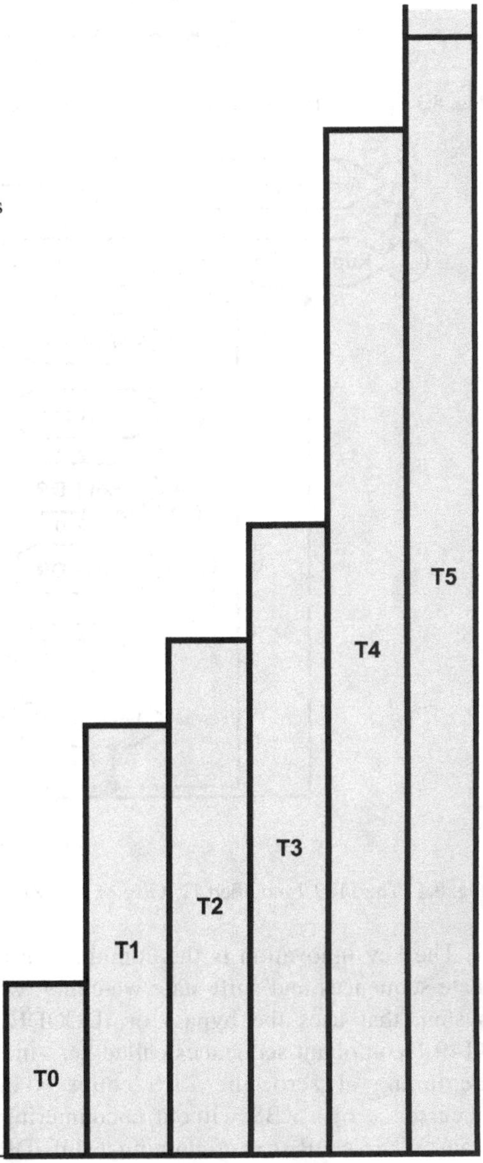

T0 T1 T2 T3 T4 T5

from the select-DR-scan TAPC state, proceed to the update-DR TAPC state without passing the shift-DR TAPC state. From the testlogic reset TAPC state, wherein the ZBS count is set to zero, the extended control mechanism is initiated when at least two ZBSs are detected before a subsequent nonzero-bit DR scan, which locks the ZBS count. A locked ZBS count of two provides access to the 1149.7 commands and registers.

Commands are typically 10-bit values and consist of two consecutive DR scans while the controller is locked at control level 2. Command part 1 (CP1) provides a 5-bit operating code, and command part 2 (CP2) provides the immediate operand, which is the lower 5 bits of the command. The function specified by the command is performed when CP2 completes (Tables 8.2 and 8.3).

A three-part command can be created by appending a third DR scan (a control register or CR scan) after CP1 and CP2 and transporting a data value. Each of the three three-part commands has a special purpose.

T1 also provides for power management through four modes of power control for the TAP. These four modes are:

1. Allow power down if TCK stops at logic one for more than 1 ms.
2. Allow power down if TCK stops at logic one for more than 1 ms in the testlogic reset TAP control state.
3. Allow power down if the device is in the testlogic reset TAP control state.
4. Do not allow power down (the test logic is always powered).

When a power-down mode is supported, the TAP is directed to resume powered operation when the run test/idle TAP control state is forced for at least 100 ms and at least 3 TCK(C) ticks.

Class T2 offers a chip-level bypass mechanism that shortens scan chains and another mechanism that provides hot connect capability. Because JTAG's serial architecture makes it complex to communicate exclusively with one specific device in the scan chain due to interactions with other devices in the chain, particularly when multiple devices or cores are combined into one chip, 1149.7 provides

Table 8.2 T1 class control levels

Control level	Overloaded function	DR scan path
0–1	None	System
2	Commands	Chip-level bypass bit
3	None (reserved)	Reserved
4–5	Auxiliary scan paths	User defined
6–7	DTS utilizes these levels	User defined

Table 8.3 TAP7 controller address

MSB		LSB			
34	27	26	11	10	00
NODE_ID[7:0]		DEVICE_ID[27:12]		DEVICE_ID[11:0]	
		Part number		Manufacturer	

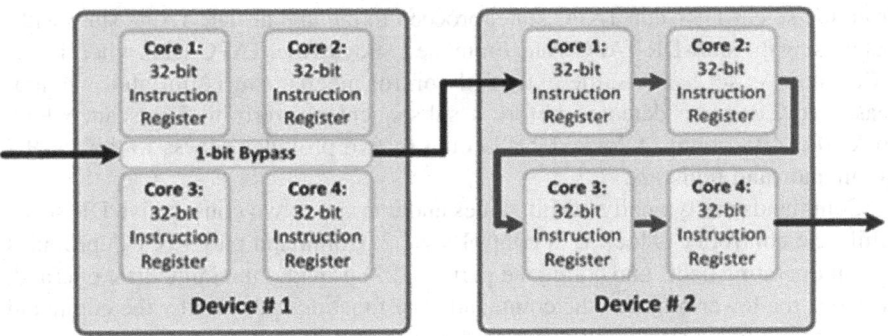

Fig. 8.5 Class T2 JSCAN2 1-bit bypass (super bypass) mode. *Source*: IEEE 1149.7

a method to address and access specific devices in the serial chain individually, without having to shift bits through the entire instruction register length of the full scan chain, as seen in Fig. 8.5.

Class T2 adds three scan formats to implement these new features:

- JSCAN0: Offers IEEE 1149.1-compliant operation.
- JSCAN1: Provides hot connection and disconnection protection. At power-up, it can select a 1-bit bypass path (also called super bypass) that is active for IR scans and DR scans. This protects TAPs from spurious signals and prevents core corruption during hot connections.
- JSCAN2: Implements the 1-bit super bypass according to the value of an 1149.7 register to improve series-connected device performance. The mechanism also functions as a firewall, enabling access to chip TAPs only after a predetermined sequence is initiated. This JScan2 provided activation/deactivation of the bypass provides a basic security that ensures that only a debug test controller can access the system once a running, powered target has a stable electrical connection.

A chip-level bypass mechanism reduces the overall scan chain length by putting unused devices in a 1-bit chip bypass mode. Using this feature can make very long scan chains dramatically shorter and improve the overall scan efficiency throughput.

Class T3 introduces the first features that are not directly extensible from the 1149.1 JTAG. Whereas classes T0–T2 continue to be based on the JTAG serial interface of data being propagated though TDI and TDO interfaces, class 3 is based on data access using a parallel interface, where each TAP has direct access to a common TDI and TDO, in addition to the common TCK and TMS signals. The 1149.7 documentation refers to parallel configurations as a star topology. Although 1149.1 references the use of parallel configurations, it does not do so in enough detail to be usable. 1149.7 provides an additional new scan format, JSCAN3, in class T3 to support star access. Figure 8.6 shows the series scan topology and the parallel interfaces of the star-4 or wide star configuration.

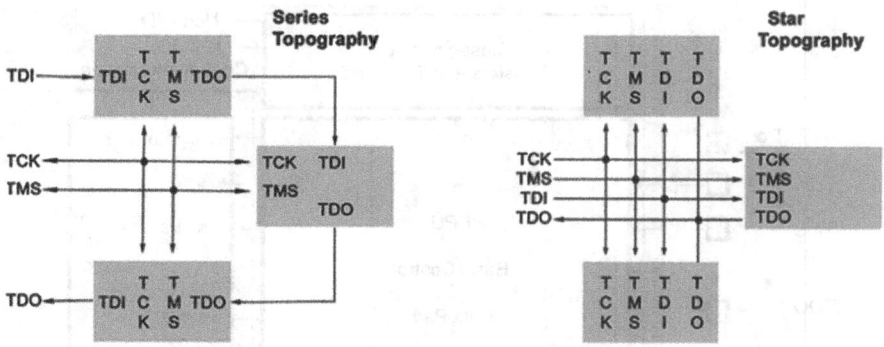

Fig. 8.6 JTAG series versus star configurations

A key provision required in a star topology with multiple TDO drivers is the prevention of drive conflict on the TDO pin. The JScan3 format is managed so that when multiple TAP.7 controllers are active, drive on the TDO pin is inhibited until an arbitrated resolution is complete.

As in a more parallel bus with multiple drivers, class T3 needs direct addressability in each of the TAP controllers. This is addressed by a TAP.7 controller address (TCA) for each of the TAPs that can drive the TSO pin. TCA values corresponding to DEVICE_ID are inherited from the 1149.1 device identification register capture value for the TAP controller. The assignment of the NODE_ID is left undefined, so that it is more easily adopted to different types of instruments and debug blocks.

The NODE_ID serves to distinguish multiple TAP.7s on a given topology branch even if they are of the same device type.

1149.7 maintains compatibility with the IEEE 1149.1 standard by making all operations appear to be series scans using Capture-xR and Update-zR TAPC states in a group of selected IEEE 1149.7-enabled TAP controllers. To operate in this mode, devices (either cores or chips) in the star configuration must be assigned controller identification (CID) numbers. An iterative arbitration system is used to assign CIDs, and operations are executed using control level 2.

The controller IDs (CID) allow the system architecture to be interrogated by external devices, such as a debug tool at connect time, and enumeration of TAPS allows specifics of what debug resources are available on the chip. This presents a significant advantage for systems with debug blocks from different sources, which otherwise would need to be known *a priori*.

Class T4 and T5 functions add new capabilities that are implemented in an advanced processing unit (APU), shown in Fig. 8.8.

Class T4 adds scan formats to support transactions with two pins instead of four, resulting in fewer total pins required on chip packages. The key to two-pin operation is eliminating the original data lines and sending bidirectional serialized data over the test mode select (TMS) line, which is renamed as TMS counter (TMSC). To implement this capability, the star configuration from class T3 is used, this time without TDI and TDO. This is the star-2 configuration, shown in Fig. 8.9.

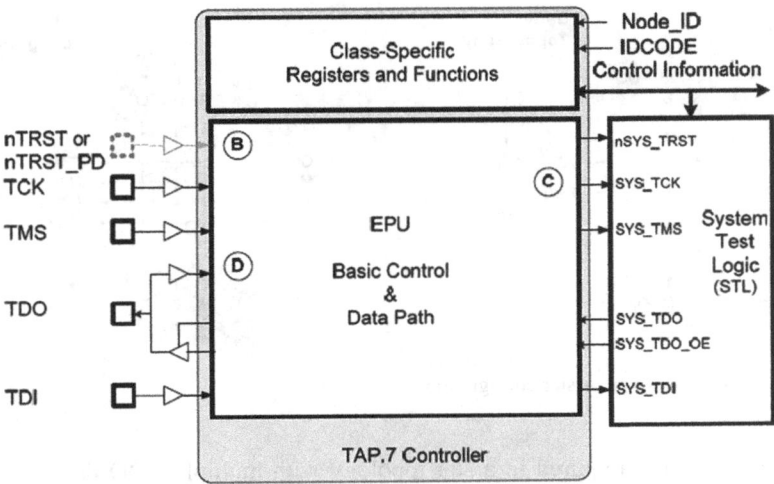

Fig. 8.7 An EPU-based TAP.7 controller for T1–T3. *Source*: IEEE 1149.7

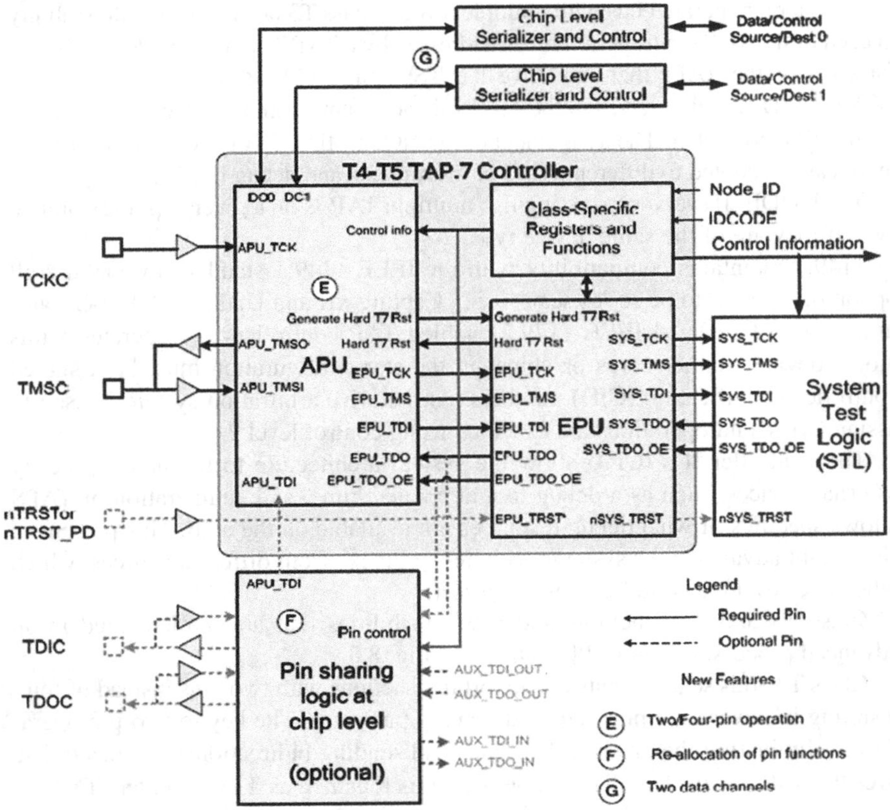

Fig. 8.8 An APU + EPU–based TAP.7 controller for T4–T5. *Source*: IEEE 1149.7

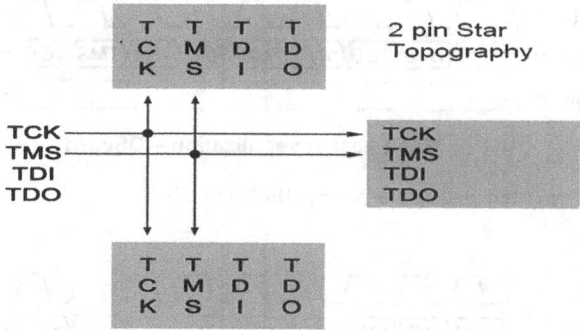

Fig. 8.9 Class T4 2-pin star configuration

In addition to reducing pin count, class T4 defines optimized download-specific scan (OSCAN) modes in which only useful information is downloaded, as shown in Figs. 8.10 and 8.11. To improve pin operation performance, the clock rate can be doubled. These features, combined with the optimized transactions, do not cause performance loss, instead improving performance in some cases.

A basic scan format that supports the advanced protocol is OSCAN1, which provides serialization of the scan packet. As shown, the TDI bit information is inverted. For each cycle in which the TDO bit appears, it is driven from the selected device in the target system back to the chip interface.

Other OSCAN formats provide optimizations in which the scan packets omit bits that carry no significant information. An example is the OSCAN7 format that is optimized for downloads from the interface to the target system. For OSCAN7, only the TDI bit information is included in the packets sent during Shift-xR TAP control states.

Class T5 adds features that improve performance and flexibility for utilizing a JTAG TAP for debugging, shown in Fig. 8.13. Whereas class T4 has primarily addressed the use of serialized packets for scan, T5 offers the capability to inter-leave transfers of nonscan data among the scan transfers. This is referred to as transport and has two variants:

- Background data transport, which uses idle bandwidth during TAP IDLE, PAUSE_DR, and PAUSE_IR for transfers.
- Custom data transport, which implements a custom link protocol to "on the fly" change direction of the data transfers.

Both types of transport can use any combination of run test/idle, Pause-xR, and Update-xR TAPC states, after which transport packets can be inserted. The distinction is that, whereas BDX has fixed allocation of I/O bandwidth available to the chip-level data channel, CDX has a custom allocation of I/O bandwidth as determined/defined by the chip-level unit (Fig. 8.14).

Class T5 gives the TAP the ability to perform debug and instrumentation operations concurrently (data is transferred during idle time), which reduces the number of

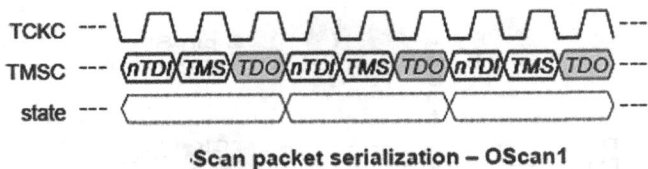

·Scan packet serialization – OScan1

Fig. 8.10 An OSCAN1 timing diagram. *Source*: IEEE 1149.7

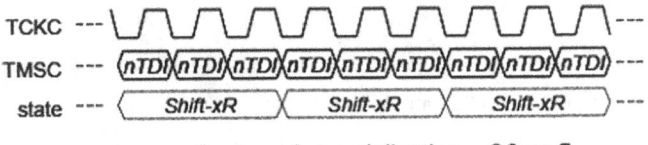

·Scan packet serialization – OScan7

Fig. 8.11 An OSCAN7 timing diagram. *Source*: IEEE 1149.7

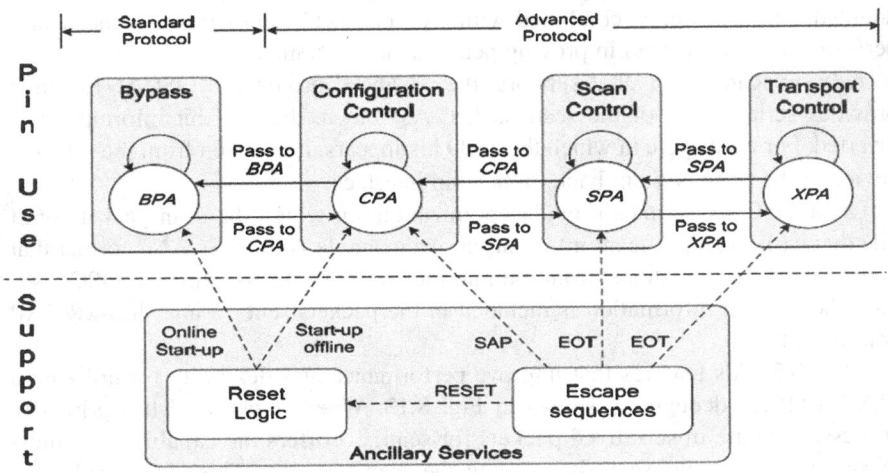

Fig. 8.12 T5 TAP.7 transitions to advanced modes. *Source*: IEEE 1149.7

pins required to address instrumentation bandwidth, and enables custom protocols to use the pins, as shown in Fig. 8.12. This is one of the attractions of 1149.7 to IEEE 5001, as discussed later. Class T5 standardizes the process to access the pins for debug as opposed to the diverse, ad hoc, and proprietary means to address interfaces supporting debug features.

Several aspects of the 1149.7 system architecture make debug instrumentation more accessible, providing for access consolidation and management of embedded TAP controllers (T0), star topology (T3), pin reduction (T4N/T5N), and capability for the TAP.7 to transport background data; custom protocols using BDX and CDX (T5) were developed with debug operations in mind.

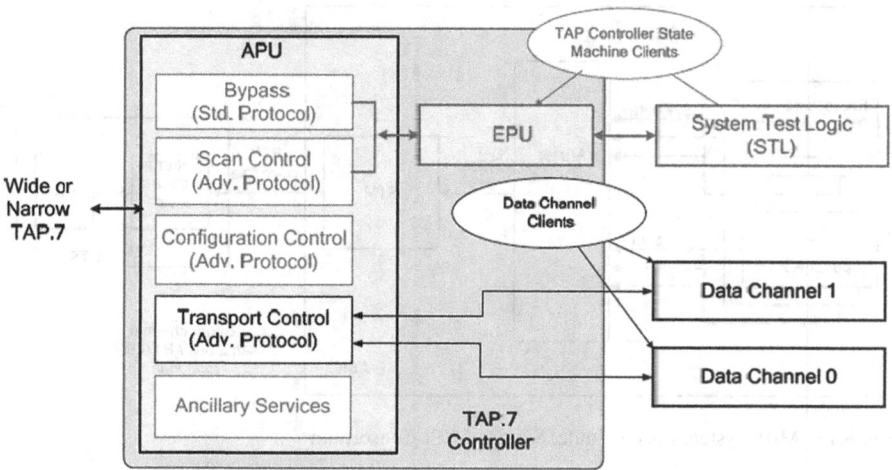

Fig. 8.13 APU functions of T5 TAP.7. *Source*: IEEE 1149.7

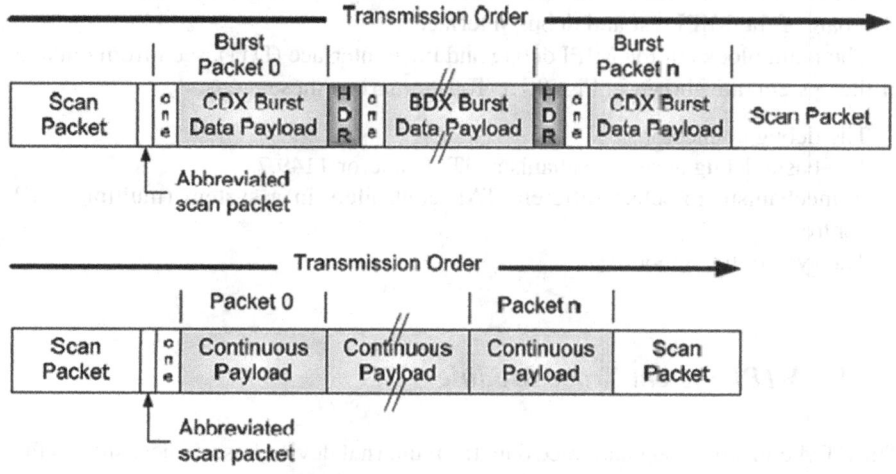

Fig. 8.14 Custom debug mode burst and continuous transfers. *Source*: IEEE 1149.7

8.3 MIPI Use of 1149.7

Mobile industry processor interface (MIPI) is an industry-standard organization addressing hardware and software interfaces found in mobile terminal systems. It maintains a test and debug working group (T&DWG) that was the original driver for 1149.7 activities. Because 1149.7 enables 2-wire pin-out options for a JTAG TAP, it is of interest to the mobile products industry. The T&DWG also specifies a system trace module (STM). STM consists of a system trace protocol (STP) and the parallel trace interface (PTI). This allows collection of debug and trace data from

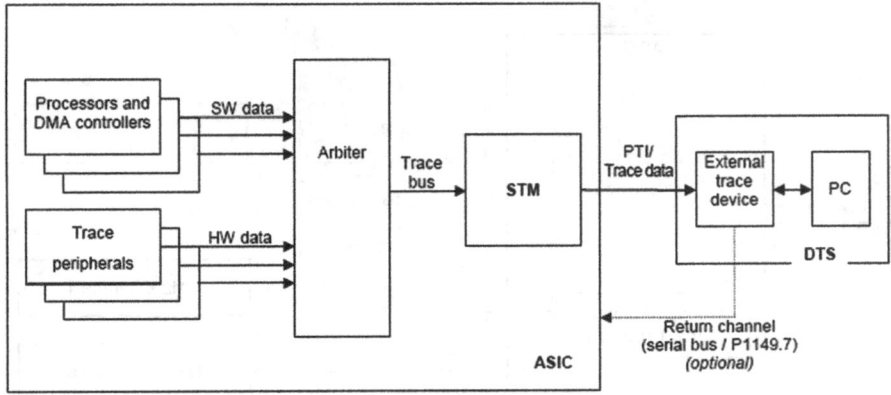

Fig. 8.15 MIPI system trace module. *Source*: MIPI Consortium

internal buses and output to an external trace capture device using a minimum set
of pins. The signals and pins required for these interfaces are given through the
"MIPI Alliance Recommendation for Test & Debug – Debug Connector," which is
also part of the MIPI test and debug interface.

The main blocks of the MIPI debug and trace interface (DTI), seen from outside
of the system, are shown in Fig. 8.15. To summarize, these are:

- The debug connector.
- The basic debug access mechanism: JTAG and/or 1149.7.
- A mechanism to select different TAP controllers in a system (multiple TAP
  control).
- The system trace module.

8.3.1 MIPI System Trace Module

The STM collects debug and trace data from internal device buses, encapsulates the
data, and sends it out to an external trace device with the following features:

- SW-generated trace optimization.
- Automatic timestamping of messages.
- Supports up to 255 HW trace sources:

  – Allows simultaneous tracing of 255 threads without interrupt disabling.

- Configurable export bus with selectable width 1/2/4-pin (+ dedicated clock
  + optional return channel):

  – Minimal pin usage is 2 pin (1 data + 1 clock).
  – Maximum pin usage is 6 pins (4 data + 1 clock + 1 return channel).

- Maximum operating frequency is 166 MHz (double data rate clocking).

- Provides a maximum bandwidth of slightly greater than 1 Gbit/s (theoretical maximum of 1.6 Gbit/s).
- Support for 8-, 16-, 32-, and 64-bit data types.

A maximum of 255 different bus masters can be connected to the STM trace port via a bus arbiter. The bus masters can be configured for either SW or HW type to optimize the system for different types of trace data.

SW-type master messages are used to transmit trace data from OS processes or tasks on 256 different channels. The different channels can be used to logically group different types of data so that one can easily filter out the data irrelevant to the ongoing debugging task. The message structures in STM are highly optimized to provide an efficient transport, especially for SW-type master data. An example of trace data output is given in Fig. 8.16.

SoCs can be designed with a 1149.7 wide interface (4 pins) or a 1149.7 narrow (2 pin) interface. As discussed previously, the 1149.7 wide devices have the normal

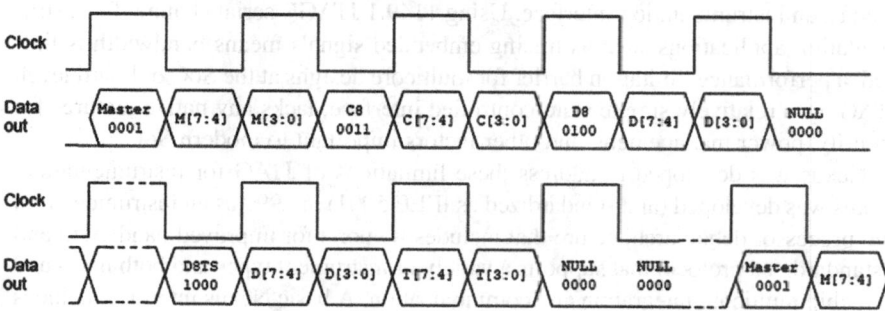

Fig. 8.16 STM output timing example

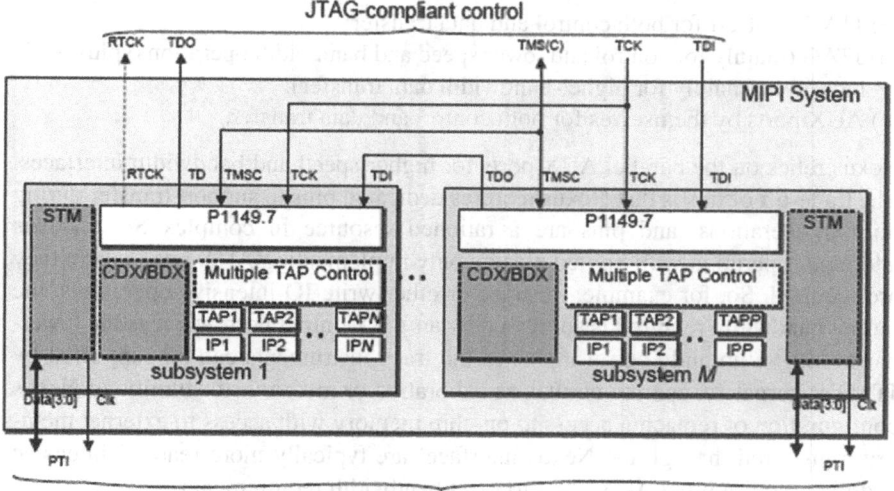

Fig. 8.17 A multi-TAP MIPI system. *Source*: MIPI Consortium

TCK, TMCS, TDI, and TDO pins and can be used as normal JTAG devices but can also be switched to advanced operating mode (2-pin protocol). Devices with a 1149.7 narrow interface only have the TCK and TMSC pins and will only carry out advanced messages (Fig. 8.17).

8.4 Nexus Use of 1149.7

Note: IEEE 5001 Nexus is discussed in depth in Chap. 11.

1149.7 also provides new capabilities for improving embedded control and visibility of chip-level analysis and design for debug logic and interfaces using IEEE 5001 (Nexus). As discussed in other chapters, Nexus provides a standard method and architecture for trace-oriented embedded instrument interfaces. 1149.1 JTAG, which has been a part of the Nexus infrastructure since its initiation, has limitations when used as an instrumentation interface. Using 1149.1 JTAG's serial channel for instrumentation applications such as tracing embedded signals means bandwidth is limited, a performance limitation barrier for multicore designs at the SoC or board level. JTAG, as a relatively simple state-controlled interface, lacks any native features for security, power management, and other factors important to modern SoC.

Nexus was developed to address these limitations of JTAG for instrumentation. Nexus was developed (and standardized as IEEE 5001) in 1999 as an instrumentation and processor debug architecture that includes IO ports for improved bandwidth and a standardized protocol that supports a variety of instrument types and both inter- and intrachip multicore integration and communication. A basic Nexus interface includes both a JTAG interface and parallel input and output data interfaces, referred to as AUX ports. Nexus interfaces can be configured in three modes:

(a) JTAG by itself for both control and data transfer.
(b) JTAG (mainly for control and lower speed and bandwidth operations) plus AUX interfaces (mainly for higher-bandwidth data transfer).
(c) AUX ports by themselves for both control and data transfer.

Nexus relies on the parallel AUX ports for higher-speed and bandwidth interfaces. The trade-off of this is that Nexus requires dedicated pins to support transfer during normal operations, and pins are a rationed resource in complex SoC. Nexus addresses operation with limited pins by only implementing AUX ports where they are required. So, for example, for trace or other write-IO-intensive operations, the output bandwidth required is addressed by an AUX output port, whereas the lower-bandwidth setup and control inputs to the trace instruments can be supported by JTAG. Alternately, operations such as calibration or memory substitution (a Nexus configuration of replacing access to on-chip memory with access to external memory transferred through the Nexus interface) are typically more read-IO intensive and may require input AUX ports to meet bandwidth requirements.

The addition of Nexus interfaces improves the instrument interface bandwidth by using the AUX ports and a higher-performance instrumentation interface

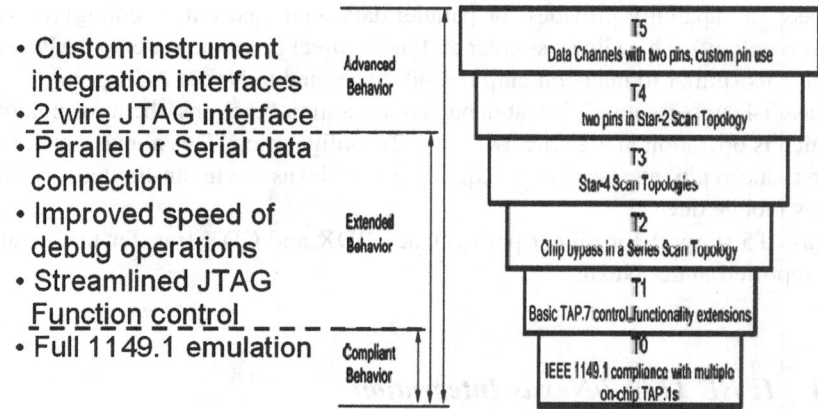

Fig. 8.18 IEEE 1149.7 class implications for Nexus

protocol. IEEE 1149.7 interfaces are backward-compatible with 1149.1 JTAG but support parallel TAP architectures that provide a more compact and powerful means for adding embedded instruments to digital processors and SoC devices. IEEE 1149.7 expands on key features of 1149.1 JTAG in several important ways. 1149.7 features are provided in a progressive series of six access port classes (T0–T5 as shown in Fig. 8.18). Support for any class of capabilities implies support for the features in the previous class. Since a Nexus packet is a message protocol, it is operating at a different level of hierarchy compared to the physical layer-oriented 1149.7 interfaces. As such, different Nexus implementations may adopt 1149.7 features up to any class level, without being required to support the entire set of features.

The costs of IEEE 1149.7 are in adding levels of control hierarchy to a test and debug port. Whereas IEEE 1149.1 JTAG was designed for reduced logic at a time when gates count per device were smaller and more expensive, in many current ICs additional logic and complexity are justified for increased features and a more flexible and reduced pin interface.

The most basic class T0 capability provides a backward-compatible 1149.1 interface. This allows legacy compatibility with existing Nexus systems or where more advanced features are not required.

Class T1 capability provides four user-selectable power control modes (based on power-down states and power-down after delay). IEEE 1149.7 selectable power mode controls may be propagated into Nexus logic to enable standardized power-down features for low-power devices.

Class T2 capability provides coupling mechanisms for reduced bypass delay for the case of multiple TAPS on a chip or a system. This allows shorter latency for serial scan operations, which can be significant for Nexus systems with several instruments.

Class T3 capability provides for parallel data input and output configurations (star-4 topology). This allows simpler and more direct access to on-chip TAPs and instruments similar to many on-chip bus interfaces and to Nexus AUX ports.

Class T4 supports multiple test/debug configuration functions, the most notable of which is operation in a 2-wire TAP (star-2) configuration. This allows operation with a reduced pin interface, a key requirement in Nexus use in pin-limited systems such as mobile devices.

Class T5 support for higher performance BDX and CDX transfer modes are not supported under Nexus.

8.4.1   IEEE 1149.7/Nexus Integration

Although IEEE 1149.7 provides new port-level features and improvements in latency for transport of embedded instrument control and data, it does not significantly address the issues of improving bandwidth in underlying communication with the instrument, a key limitation in performance of applications such as trace and calibration. Even at its most advanced (T5) level, the IEEE 1149.7 interface only defines a single data pin for instrument use. Having a scalable and extendable data interface for test and debug was one of the drivers of Nexus, and one of its most notable features is the definition of input and output AUX ports that increase the effective number of data channels to significantly increase transport bandwidth.

By defining parallel interfaces, 1149.7.7 star-4 configurations can be synchronized and integrated with both AUX IN and AUX OUT ports (Fig. 8.19). Depending on the requirements, only one (IN or OUT) of the AUX ports may be required. Number and size of the AUX ports are configurable and typically trade off with output buffer size. For a multi-TAP configuration, larger or additional buffers can be required because the synchronized AUX port for each TAP may be stalled by activity on other TAPs or by the 1149.7 interface itself. In addition, support for advanced modes, CDX in particular, may require that TAPs communicate and arbitrate access to the channel. A similar configuration can be used to implement Nexus with a 2-wire star-2 configuration. IEEE 5001 also defines, in the 2009 revision, the ability to implement increased (order-of-magnitude) bandwidth by replacing the AUX channels with a corresponding number of SerDes ports allowing gigabit data transfers.

Nexus packet-based transfers include ID fields and specific transfer operations for accessing and assigning ID information that allows sequential and multiplexed access to multiple instruments within a single Nexus interface or across multiple Nexus instantiations.

In a typical mode of operation, a Nexus interface may consist of a lower-bandwidth JTAG interface providing command and configuration inputs and a higher-bandwidth AUX OUT port outputting debug data. Because the JTAG chain

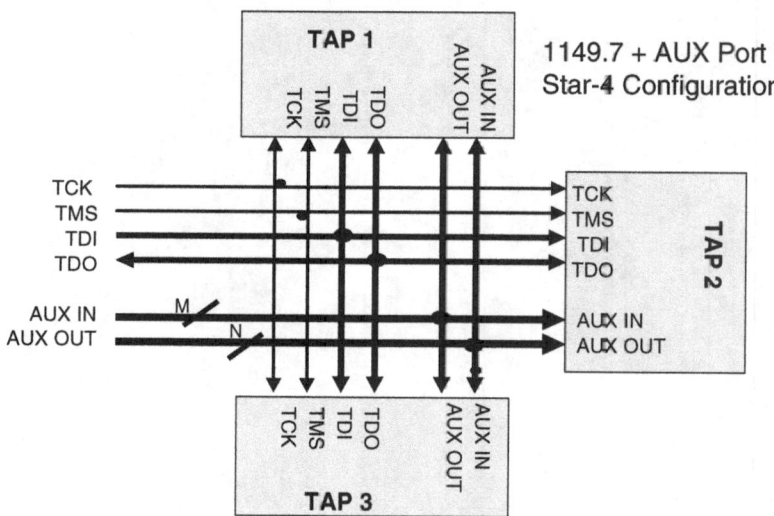

Fig. 8.19 IEEE 1149.7 star-4 and Nexus AUX port interfaces

may have greater latency than the AUX ports, this can in some cases introduce communication complexity and reduced performance. IEEE 1149.7, in particular in the T3 and T4 star configurations, can reduce this latency. The AUX ports, as parallel data channels, can be configured in a related star output interface to synchronize operations and signaling.

In summary, Nexus has adopted IEEE 1149.7 as a successor to the 1149.1 JTAG interface port that is an interface feature in the IEEE 5001-2003 Nexus specification release. Because 1149.7 is backward compatible to 1149.1 JTAG, it does not impact any legacy systems that use JTAG as a debug interface. Its value for Nexus is in new features that make debug more efficient, including:

- The ability to quickly access a specific TAP in a system with multiple TAPs, either on chip or in different devices. By implementing a system-level bypass, the scan chain is drastically shorter, which directly improves debugging performance.
- The ability to control debug logic power consumption in an industry standard manner.
- The introduction of star-4 (4-wire JTAG parallel interface to multiple TAPs) connectivity to complement the 1149.1 JAG serial TAP connections. A star configuration allows simpler test connection and simplified physical connections that are compatible with Nexus data interfaces.
- The 2-wire TAP option (star-2) that replaces the four-wire TAP to reduce pin cost.

Chapter 9
IEEE P1687 – IJTAG

P1687 (also informally but commonly referred to as IJTAG) is a set of emerging standard definitions that extend the JTAG functionality to include a variety of instruments. The scope of P1687 is not limited to debug purposes and as such it seeks to be more inclusive in the functions it defines. Whereas other instrumentation-related standards efforts, such as the IEEE 1149.7 standard, define a next-generation TAP and pin interface while maintaining backward compatibility with the IEEE 1149.1, P1687 is an effort to standardize connection and communication with on-chip instruments for the control and management of embedded instrumentation within a semiconductor device while retaining the 1149.1 TAP interface.

P1687 efforts started in 2004, and the IEEE PAR (project authorization request) was approved in 2005. The P1687 standard addresses descriptions of how to connect both JTAG and non-JTAG on-chip instruments and to define (in addition to BDSL) languages for communicating with the instruments via 1149.1 test data registers. The languages that potentially are being selected to support P1687 are:

ICL – Instrument Connectivity Language, which adds information above BSDL (including operation of the instrument).

PDL – Procedural Description Language, which adds reusable vectors.

TCL – Tool Command Language, for scripting debug applications built with the TK toolkit.

IJTAG operations use the concept of a compliant 1149.1 overlap zone to define compatibility and to ensure that IJTAG instruments do not violate any rules associated with JTAG operations.

As an overview of P1687, the comprehensive Fig. 9.1 shows the scope of instruments that P1687 seeks to address. The charter for the P1687 development effort makes clear their intent in preserving full compatibility with 1149.1 JTAG and that all operations that require it use an 1149.1-compliant TAP and TAP controller. In particular, the guidelines are as follows:

1. 1149.1 does not require any compliance-enable mode to use or access the 1687 portion of the architecture.
2. A device supporting 1687 can be intermingled with other (traditional 1149.1) devices in a multiple 1149.1 board-test system.

N. Stollon, *On-Chip Instrumentation: Design and Debug for Systems on Chip*, DOI 10.1007/978-1-4419-7563-8_9, © Springer Science+Business Media, LLC 2011

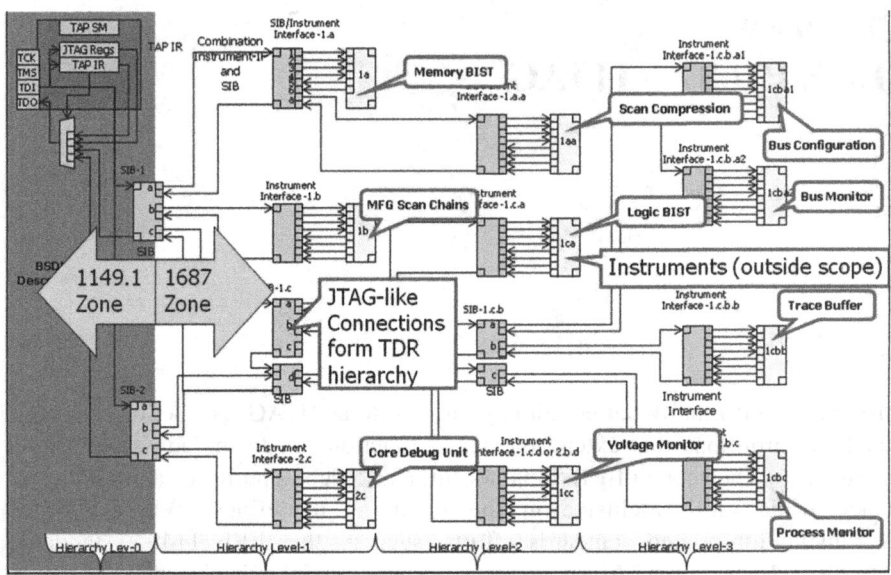

Fig. 9.1 The P1687 universe. *Source*: ASSET InterTech, Inc. All rights reserved

3. No special functions, logic, or filters are required in front of the 1149.1 TAP (external to the device or between the 1149.1 TAP and the TAP controller) for 1687 purposes.

4. P1687 does not chance any of the public 1149.1 standard features, including the 1149.1 TAP; the 1149.1 state machine, defined registers (the boundary scan register; the bypass register; the ID code register; the instruction register; etc.) or the public instructions (EXTEST, INTEST, SAMPLE, PRELOAD, IDCODE, CLAMP, HIGHZ, etc.).

5. BSDL is used as a preliminary checker for instruments included within the 1149.1 zone that may be accessed with declared public instructions or private instructions. If it can not be described in BSDL, it should not be in the 1149.1 zone.

6. The architecture defines an overlap zone shared by 1149.1 and 1687 elements, which is 1149.1 compliant and can be described using BSDL. BSDL compatibility ensured that connection schemes will be driven by 1149.1 requirements and compliant with 1149.1 criteria.

7. 1687 does not replace or modify interface or port elements defined in the 1149.1 standard and references the JTAG TAP, as the port, controller, and access point to the 1687 gateway.

8. Adding 1687 features and functions should not impact operation or use of 1149.1 complaint portions of the architecture.

9. The 1149.1 overlap zone, which is a logic portion of containing both 1149.1 and 1687 hardware, places 1149.1-compatible instruments and/or 1149.1-defined test data registers and/or hierarchy-support elements within parts of the 1149.1 overlap zone that can be described by BSDL.

10. Gateway elements are defined as instruments that enable hierarchical access (access to other instruments that do not require a direct IRScan).
11. User-defined instructions to the 1149.1 instruction set provide control to both JTAG and 1687 gateway elements contained within the 1149.1 overlap zone.
12. The connectivity of instruments in the 1149.1 zone should be driven by 1149.1 methods and be compatible with 1149.1 operations.
13. 1687 instruments that are not compatible with 1149.1 (i.e. cannot be described by BSDL) should not be directly connected to the 1149.1 IR and should not be in the 1149.1 overlap zone (but rather should be moved to a dedicated 1687 zone).
14. 1687 instruments in the 1149.1 overlap zone should be the only non-1149.1 logic that can react to IR-Scan operations; all other 1687 instruments should be accessed, configured, and controlled using only DR-Scan (shift-DR and the update-DR) operations.

9.1 Overlap Zones and Gateway Elements

Two key concepts of P1687 are those of 1149.1 overlap zone and gateway elements (or gateway instruments). Referring to Fig. 9.2, the 1149.1 overlap zone can be seen with the 1149.1 portions shown on the left and the 1687-only portion on the right of the line (the 1149.1 overlap zone) bisecting Fig. 9.2. Gateway elements that

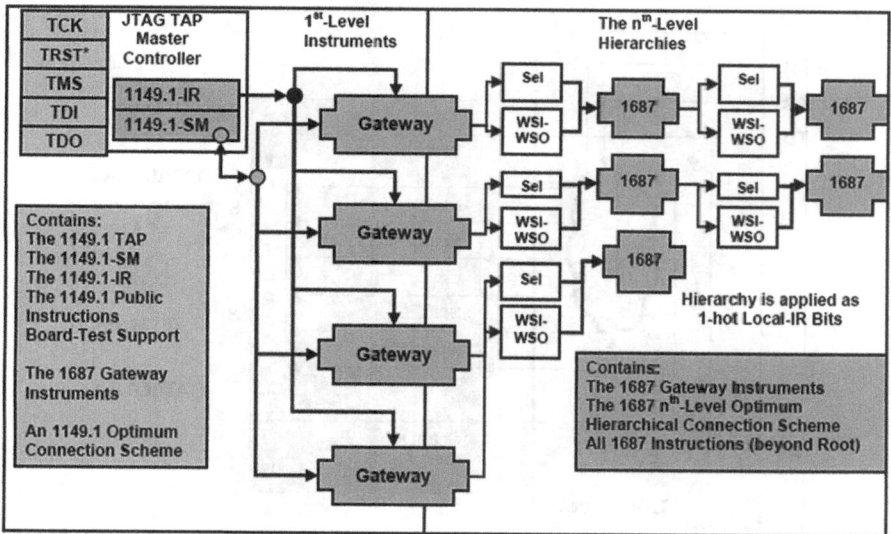

Fig. 9.2 Overlap zone with gateway elements. *Source*: ASSET InterTech, Inc. All rights reserved

straddle this zone are the only logic overlapping the two zones. The gateway element supports hierarchical connection to other instruments that are described in the 1687 zone but also needs to be fully compatible with the 1149.1.

All instruments in the 1149.1 portion of the architecture, including the gateway elements need to be compliant with the 1149.1 TAP FSM, Operations are enabled by IR-Scans that load instructions in the 1149.1-IR and are active on the falling edge of TCK in the update-IR state. One criteria that P1867 uses to determine if an element belongs in the 1149.1 zone is if it be described by BSDL. If it can not, it is not 11491 compatible.

The 1687 portion of the architecture begins at the gateway element or instrument and in turn enables other instruments to be accessed (by creating a select signal) and allows access to TDI-TDO data from the 1149.1 TAP to the target instrument in the 1687 zone. 1687-zone instruments are accessed, controlled, and configured only by DR-Scans (Shift-DR and Update-DR assertions) through a gateway element/instrument.

Figure 9.3 shows a simple generic example of a gateway element, where the test data register (TDR) receiving data through the TDI and outputting data through the TDO is connected to the 1149.1-IR and through encoding of an instruction operation can be used to pass hierarchical connections to 1687 instruments. Note that whereas the 1149.1 side of the TDR is a TJAG serial data interface, on its 1687 side it communicates as a parallel register.

TDRs are constructed of registered elements called select instrument bits (SIB); an example is a TDR bit that includes a hierarchical interface port (HIP) that enables a hierarchical connection. The hierarchical connection allows the SIB to be

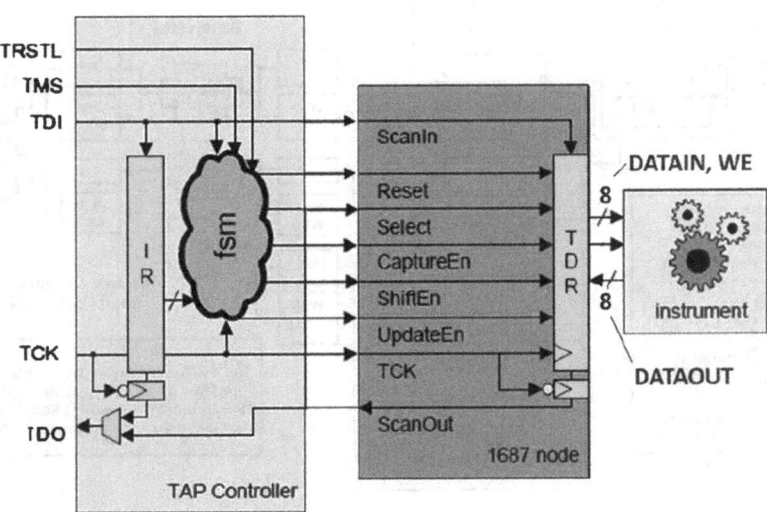

Fig. 9.3 A basic gateway element using a test data register

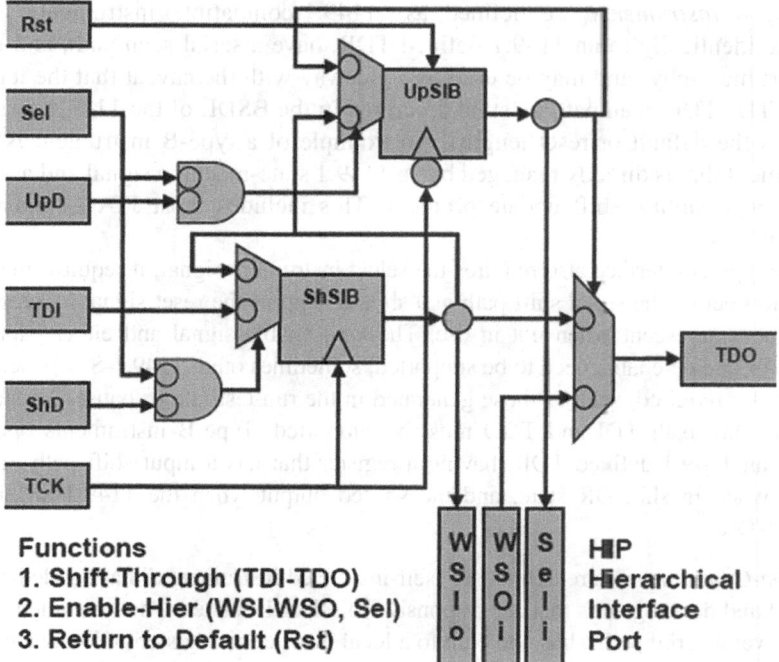

Functions
1. **Shift-Through (TDI-TDO)**
2. **Enable-Hier (WSI-WSO, Sel)**
3. **Return to Default (Rst)**

WSIo WSOi Seli   HIP
              Hierarchical
              Interface
              Port

Fig. 9.4 A select instrument bit logic block. *Source*: ASSET InterTech, Inc.

used to bypass portions of a JTAG scan chain and otherwise control the flow of the DTI/TDO path (Fig. 9.4).

9.2 Classes of P1687 Instruments

The taxonomy of P1687 defines support for four classes of instrument types labeled A, B, C, and D.

Type-A instruments are defined as "self-contained instruments" that are enabled by static signals and report status by latched-output signals. Type-A instruments are not compatible to the 1149.1 interface. Typically, they don't have interfaces for serial paths (and therefore, cannot be used as a gateway). An example of a type-A instrument is a simple memory BIST controller.

Type-A interfaces require a select instrument signal and is controlled by decoded instruction bits from an instruction register or other status signals. The type-A instrument interface typically used on a "self-contained" instruments such as logic BISTs, memory BISTs, and other stand-alone-instruments. Communication is with static signals that are applied on the 1149.1 update-DR and sampled by the 1149.1 capture-DR.

Type-B instruments are defined as "1149.1-compatible instruments" that operate identically to an 1149.1-defined TDR, have a serial-scan path, and may support hierarchy (and may be used as a gateway with the caveat that the length of the TDI-TDO scan path must be described in the BSDL of the 1149.1 overlap zone as the default or reset length). An example of a type-B instrument is any instrument that is directly managed by an 1149.1 state-machine signal and associated select capture shift update protocol. This includes most JTAG-controlled trace blocks.

The type-B interface also requires the select instrument signal; it requires the test clock to operate the serial shift path and should support the reset signal to keep the instrument quiescent when not in use. The shift-enable signal and either capture-enable or update-enable need to be supported; sometimes other 1149.1-SM-generated signals are required, such as those generated in the run test/idle or pause-DR states, and the shift-path TDI and TDO must be supported. Type-B instruments operate simlar an 1149.1-defined TDR, having a register that has a input shift path that is active when in shift-DR state; and the shifted output when the 1149.1-SM is in capture-DR.

Type-C instruments are defined as "self-instructed instruments," having dedicated control and data registers that are responsive to select-IR states. The instrument may have several serial paths, has one path to a local instruction register and may support hierarchy. Type C instruments may be used as a gateway. An example of a type-C instrument includes most processor debug blocks (including ETM and Nexus).

The type-C interface also requires the select-instrument signal, requires the test clock to operate the selected serial shift path, and should support the reset signal to keep the instrument quiescent when not in use; the shift-enable and update-enable signals and optionally the capture-enable need to be supported; sometimes other 1149.1-SM-generated signals are required such as those generated in the run test/ idle or pause-DR states; and the serial shift-path TDI and TDO must be supported. The type-C instrument is one that operates like a 1500-defined test access mechanism (TAM) – a set of registers, one of which is active as a shift path that is active when the 1149.1-SM is in shift-DR and is based on the instructions in a register defined as an instruction register.

Type-D instruments are defined as type-B or type-C instruments whose control interface supports at least one of the following: a signal or sequence not produced by a compliant 1149.1 TAP or 1149.1 controller; a clock other than TCK; or a data port other than the TDI-TDO serial scan path (and hence cannot be used as a gateway because it is not easily described in BSDL but it may still be used as a hierarchical instrument). An example of a type-D instrument is a 1500-wrapped core with core boundary scan cells that require the transfer signal.

The type-D interface can be identical to the type-B or type-C but must have at least one non-1149.1-compatibility issue, such as the instrument requiring:

• A clock in addition to or other than test clock (the 1149.1 TCK) to operate some portion of the interface.

- A data path other than the defined TDI-TDO serial shift path that synchronizes to the test clock (such as a parallel port that synchronizes to a test clock or a parallel port that synchronizes to an alternate clock or a high-speed serial port that synchronizes to a high-speed alternate clock).
- A signal not provided from a compliant 1149.1 controller such as stall, bus request, data valid, and counter done).
- A sequence not provided by a compliant 1149.1 controller.

Type-D instruments are expected to be instruments such as a bus controller, bus converter, or clock controller that can be configured and controlled through the 1687 architecture.

9.3   IEEE 1500 Instruments

P1687 applications occasionally refer to 1500-wrapped instruments, so we discuss these briefly in the interest of comprehensiveness. See IEEE Standard 1500, the IEEE standard for embedded core, for a more complete discussion.

The IEEE 1500 standard defines a mechanism for the test of cores within a system on chip, including a wrapper hardware architecture. It also uses a core test language (CTL) to facilitate communication between core designers and integrators. IEEE 1500 defines standard components and general wrapper architecture, including wrapper parallel input and output ports, core functional inputs and

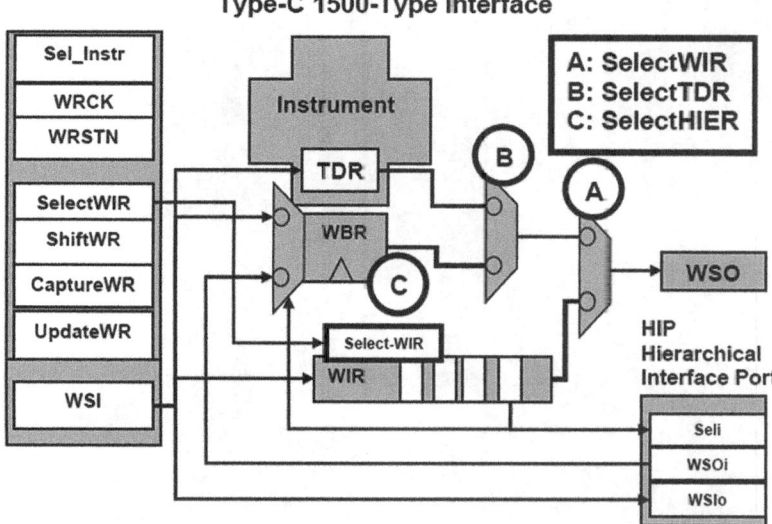

Fig. 9.5 An example 1500-type instrument interface. *Source*: ASSET InterTech, Inc. All rights reserved

outputs, wrapper serial input, and serial output for test (see IEEE Standard 1500–2004) (Fig. 9.5).

An example of a 1500 test access mechanism is one in which the connectivity and protocol structure used to access an instrument gateway is enabled when one or more of the instructions in the 1500-defined wrapper instruction register (WIR) can enable one or more hierarchical interface ports (HIPs).

Chapter 10
OCP IP Debug Interfaces

The Open Core Protocol-International Partnership (OCP-IP) is an industry-based standard group defining vendor-neutral socket interfaces for interconnecting cores and other on-chip components. The OCP-IP socket-based integration strategy has been proven in a multitude of designs from leading chip vendors. OCP-IP's strategic focus is deeper than other socket-based interfaces, as it address a wide range of related topics including system-level design and debug.

One concern of particular importance in OCP-IP is the support of Multicore and Multiprocessor SoC architectures. The complexities of Multicore SoC are far more than the sum or the parts, since issues can appear not just in the processors, but in their interactions. This, in turn, drove a need for instrumentation types and standardized interfaces and debug methods that would support these complex OCP socket based systems. The OCP-IP chartered a debug working group in 2005 with the specific goals of developing on-chip debug solutions that address the issues of debugging multicore systems, in which different cores may have associated debug blocks with different features, different signaling, different analysis requirements, and so on. The evolution of instrumentation interfaces to various core have led to widely different signal interfaces. The instrumentation interfaces defined by the debug working group were focused on providing a superset of signal interfaces that covered these generally incompatible debug interfaces, as well as by the need to have debug operations interacting with instrumentation for the OCP bus architecture itself.

The value of OCP-IP and debug work designed by a neutral party is that it is designed to be vendor-neutral. The competitive nature of different core providers have limited their collaboration in developing common instrumentation interfaces and methods for analysis needed for complex SoCs. Although groups such as Nexus 5001 (Chap. 11) address a vendor-neutral debug interface, this interface is primarily focused on the top-level interface (JTAG versus trace port), the protocol, and defined registers and other issues not directly related to implementing instrumentation systems at the core level.

The OCP- IP, recognizing that other groups were addressing the system-level debug interface, focused primarily on the low-level signal interfaces that are required to create a debug socket that would work with other sockets defined by OCP-IP. A common set of standard instrumentation signals creates a basis for more interactive instrumentation which can be applied to cores and architectures from

N. Stollon, *On-Chip Instrumentation: Design and Debug for Systems on Chip*, DOI 10.1007/978-1-4419-7563-8_10, © Springer Science+Business Media, LLC 2011

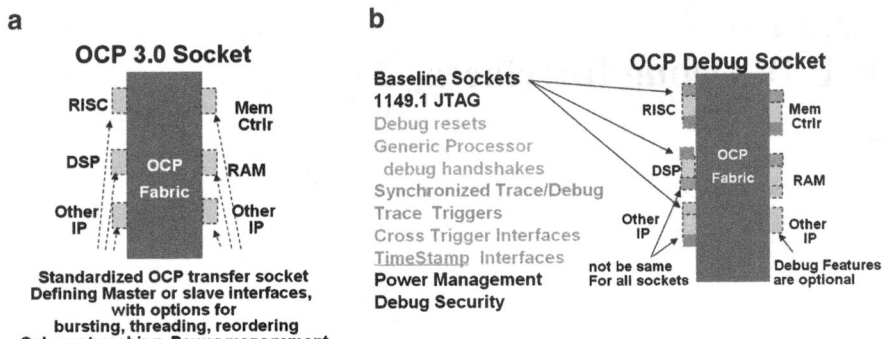

Fig. 10.1 (a) OCP sockets for various cores; (b) OCP debug interface sockets

different vendors. The advantage of common instrument interfaces becomes more important with complex SoCs with diverse architectures and heterogeneous cores and IP from different sources.

The OCP-IP has defined a set of architectures based on the idea of sockets. The OCP socket defines data and instruction signal interfaces for a processor, which allows a vendor-neutral bus fabric to be used for routing and bus connections (bus sockets are also discussed in Chap. 6). The OCP debug socket defined by the OCP-IP debug working group defines a complementary baseline set of debug functions that could be implemented on chip to include global run control signals, as well as trace, triggering, time-stamping, and other on-chip analysis functions. Increasingly, SoCs incorporate asynchronous domains, diverse voltage islands, various power-saving schemes, varying levels of embedded security, and so on; as shown in Fig. 10.1, all of which add to the complexity of the debug solution.

The OCP debug socket defines both critical and optional sets of debug signals that allow different IP blocks to communicate and coordinate their specific debug requirements and features. The baseline signals are typically common to all the IP blocks and therefore are known to be supported by a common JTAG chain or debug port. Optional signals are those supporting functions specific to particular blocks or asynchronous/secured/powered subsystems.

10.1 OCP Multicore Debug

A multicore debug interface must provide a set of signals for basic debug capabilities. These basic signals can be divided into four groups:

1. Debug control: Defines independent reset and debug-enable signals.
2. JTAG interface: Defines signals for the JTAG protocol.
3. Debugger interface: Defines a set of debug interfaces that address system-level debug of run control and debugger tool interfaces.
4. Cross-trigger interface: Defines signals for distribution of debug events and for system-level control in a multicore SoC.

A mechanism for the systemwide debug of a heterogeneous SoC uses a standardized OCP bus interface for all IP blocks. The standardized bus interface or socket provides a well-defined set of vendor-neutral signals that address interfaces between a core and the bus fabric. Specific types of debug features targeted by the debug socket(s) are:

- Signal-level observation (bus and system trace) and control (triggering).
- Consistent (multiple) processor software debugger and bus traffic observation interfaces.
- Special debug features for security islands, voltage islands, gated clock islands, and so on.
- New classes of debug errors (which are different from system errors).

The debug concepts addressed can be applied to single-core debugging (without cross-triggers, trace, or timestamping) and can be extended to more cores and channels of debug for more complex systems. For multicore chips, there is an implicit debugging requirement to observe activity of (at least) two cores in order to enable a comparative analysis of operations and communications. We use a dual-channel synchronous debug socket as a baseline model. Dual-channel debug is a minimum to enable comparison of instruction or other cause and events or other effects that occur in different cores. A means of synchronous is needed to allow these instructions and events to be displayed in correct temporal relations, One such means is by use of timestamping during collection of trace information. The idea is similar to a dual-channel logic analyzer; when cores are not in debug mode, then any two IP blocks can be observed or traced in temporal comparison with a common and extensible set of signal interfaces.

Debug Components and IP Interfaces. The basic signals for an OCP debug interface socket may be added to all cores and IP blocks that support or need debugging access. OCP debug port sockets may be implemented independently of data sockets, including at different points in the OCP system from where a data port may be implemented, as shown in Fig. 10.2.

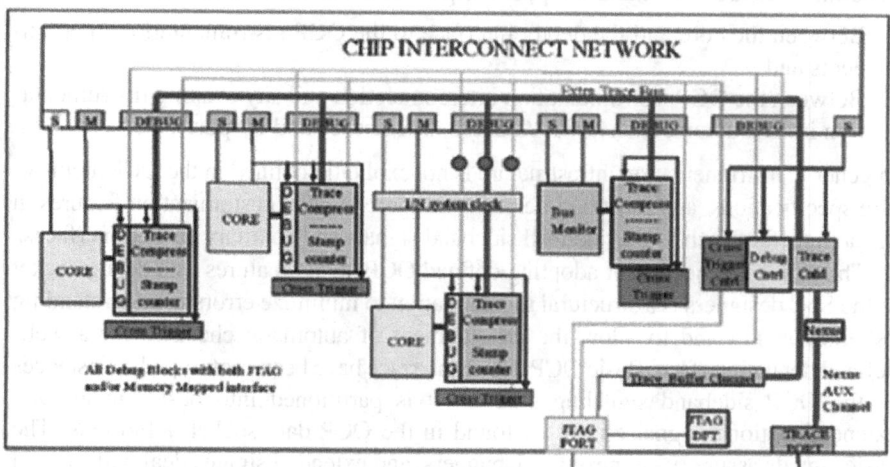

Fig. 10.2 A multicore debug socket implementation. *Source*: OCP IP

The debug signals may be implemented as a separate socket or as additional signals to the OCP data (master and/or slave) port (debug blocks are memory mapped and controlled through the OCP data socket) or as an independent OCP port configuration that can be controlled via JTAG or another external interface.

In general, although debug socket functions are passive and not intrusive to system operations or performance, some debug-related operations (such as cross-triggering) may interact with other parts of the system.

Debug Interface Definitions. The programming of registers that contain either configuration or status information in the debug IP blocks may be JTAG-mapped or memory-mapped. Either mode of control and access is acceptable, based on specific system requirements as shown by example in Fig. 10.2.

There are two preferred methods of mapping the registers of the debug IP-blocks such that all debug registers should be memory-mapped to fit into the usual programmer's models and allow for standard and extended testability concepts in manufacturing:

(a) memory space mapping: The on-chip processor core can operate the debug blocks.
(b) JTAG-mapped register access: This is controlled by external software debuggers over JTAG and can operate all debug IP blocks.

Comparative two-channel debugging with true time display of events is similar to the logic analyzer philosophy. The time-aligned display of system bus traces of data events from different initiators on different buses is the main source of information. Setting triggers on any signal or combination of events from different cores, IP blocks, and firing assertions is also basic to this idea. That is accomplished by the cross-trigger debug hardware block.

The OCP debug interface follows the general concept of master-slave request-response philosophy to provide straightforward mapping of existing signaling schemes to corresponding instrument interfaces in various cores and IP blocks. In general, there are two signal wrappers required on the hardware side:

1. Between the core and the debug interface to the OCP instrumentation interconnects and
2. Between the OCP instrumentation interconnection and any existing instrumentation infrastructure, such as JTAG TAPs, trace buses and IO ports, etc.

In general, instrumentation infrastructure is not explicitly defined in the OCP architecture specifications, and must be inferred from more general customization features in the architecture such as user defined sideband signals and auxiliary socket interfaces.

The primary objective of adopting defined OCP signal features as a debug socket to the SoC designers is a structural regularization to minimize errors in understanding its functionality and to allow the development of automatic checkers for a well-defined debug interface. Basic OCP debug interface have been prototyped as instances of the OCP sideband signaling scheme. It is partitioned into basic signals and extended optional signal groups as found in the OCP data socket definitions. The basic signals assure run control for debuggers, and extended signals deal with special situations like voltage islands, security islands, and power-down modes. Performance metering and assertions are also part of the optional signals in the debug socket.

10.2 OCP Debug Features

- Debugging in the real target system: No mechanical or electrical constraints.
- Full visibility: Cycle-accurate trace of multiprocessor, multibus SoCs.
- No limitation for low-pin-count, high-frequency devices.
- Complex triggering modes; to allow for example, negative triggering on an event not occurring in a timeframe expected to minimize amounts of collection of trace data.
- Support for code profiling and performance analysis through programmable event counters.
- Portability: The OCP debug interface is adaptable to any processor or bus architecture; software developers continue to use tools they are familiar with.
- Low cost: No expensive hardware needed to access OCP debug systems.
- Proven implementation: The OCP debug system prototype was executed successfully.
- Nonintrusive debugging of embedded multiprocessor systems.
- Target system runs at full speed in the application environment.
- Access to internal buses.
- Real-time, cycle-accurate tracing.
- Trace capabilities for:

  - Processors: Process ID, program, data, status, watch point.
  - Buses: Data, status, watch point.
  - Signals: Status.

- Complex trigger system including cross-target triggers.
- Translates raw data into meaningful messages.
- Compresses trace messages to save memory.
- Trace memory can be configured as a circular buffer to collect trace messages either continuously or before and/or after a watch point occurs.
- Implementation can be partitioned for easy adaptation to new cores.
- Instrument interfaces are not limited to a particular physical interface between chip and debug host. JTAG is a reference interface, due to proven use in instrumentation, however other available interfaces (I2C, UARTS, etc) may be used as appropriate.
- Security: The OCP debug system can be locked by default and can only be unlocked by system hardware.

As for the OCP data socket, there is a superset for the different bus interfaces and data structures and we seek to define an OCP debug socket that can be a superset of the debug solutions. Most concepts discussed are based on common denominators for the past and present debug concepts. This enables OCP compliant creation of standardized IP blocks for debug situations and purposes, including:

- Signal-level observation (bus and system trace) and control (triggering).
- Consistent (multiple) processor software debugger and bus traffic observation interfaces (GUI).
- Special debug features for security islands, voltage islands, gated clock islands, and so on.
- New classes of debug errors (different from system errors) analysis.

The debug concepts addressed can be applied to single-core debugging (without cross-triggers, trace, or timestamping), and it can be extended to more cores and channels of debug for more complex systems. For multicore chips, there is an implicit debugging requirement to observe activity of (at least) two cores out of many in order to enable comparative analysis of operations and communications. As a default, OCP debug interfaces should support multiple cores. We use a dual-channel synchronous debug socket as an example. The intention is similar to that of a dual-channel logic analyzer, and when cores are not in debug mode, then any two IP blocks can be observed or traced in temporal comparison with a common and extensible set of signal interfaces. We avoid defining a separate debug bus to keep a simple modular IP structure on the chip.

The purpose to debug in a chip can be very different. At least three variants need to be satisfied by a standard:

- Pure software debugging concentrates on minimum additions to proven hardware still providing a rich debug environment for development of software.
- Pure hardware debugging concentrates on the simplest additions in hardware to expose chip internal signals on the pins (JTAG) to be used to prove correct functionality and correct design.
- System debugging concentrates on software debug and hardware observation.

10.3   Three Views of Debugging

As a process, debug can differ between companies, projects, and points in the design cycle.

10.3.1   Pure Software Debugging

Pure software debugging concentrates on minimal additions for instrumentation to proven hardware and IP while still providing a rich debug environment for software. The debugger connects to the processor that programs all debug hardware over the system bus. Target system hardware is fully utilized for debugging. The assumption is that all hardware is correct. Special instructions and signals to let the processor prevail in locked situations are desirable and included in the basic OCP debug interface signals. This style of debugging is well documented on several chip architectures. Systems are built by connecting several proven chips together; therefore debugging with interchip cross-trigger is a second special requirement. To simplify a dual-trace memory, one trace buffer can be used in connection with a "synchronous run mode" triggered from a debug event or tracepoint signal. For short trace durations (depending on the same of the memory) this will make the ordering of events in the trace buffer in correct temporal relation possible without timestamping.

10.3.2 Pure Hardware Debugging

Pure hardware debugging concentrates on the simplest additions in hardware to expose chip internal signals on the pins (JTAG or other) or in registers to be used to prove correct internal functionality or design parameters in mission-critical applications and warranty cases. Most important for this concept is an independent clock from outside that is reliable even if the system clock is stopped. Also, triggering precise to one system-clock cycle, or local-clock cycle, is essential to let this debug hardware react exactly like assertions in a simulation. Often, signals inside IP blocks are observed. No software debuggers need to be involved in the display of this information, but we believe there are analysis advantages to including the display of such extra information.

10.3.3 System-on-Chip Debugging

System-on-chip debugging concentrates on software tracing and hardware observation requirements common in initial SoCs. Observation of the on-chip hardware interaction is essential to complete software application and verification. Comparative debugging of any two cores is equally important for multicore systems. The debug-system is independent of the target hardware and captures both "pre-reset" and "post-crash" events as well as bus traffic bottlenecks. Debugging must proceed even if the major components of the system are in power-down or a core is in sleep mode. The debug hardware may be shut down during normal chip operation for security or power improvement. In some systems, security can be enforced by making debug hardware inoperable in production chips by burning fuses or otherwise permanently disabling the instrument logic. Such a concept of debugging is best suited to support ASIC designs. To simplify a dual-trace memory, one physical buffer can be used that holds two compressed trace streams with origin tags and timestamps.

Debug features need to support the system-level verification and analysis of OCP-based systems. System level models of Instrumentation blocks should be available for EDA analysis for JTAG and DFT, BIST, and other debug structures, even when these are implemented as physical (post-synthesis gate-level insertion) macros. From a system point of view, debug blocks should support the same level of model abstractions used in other areas of a design, in order to support it with miscellaneous simulators and software debuggers and to simplify hardware analysis.

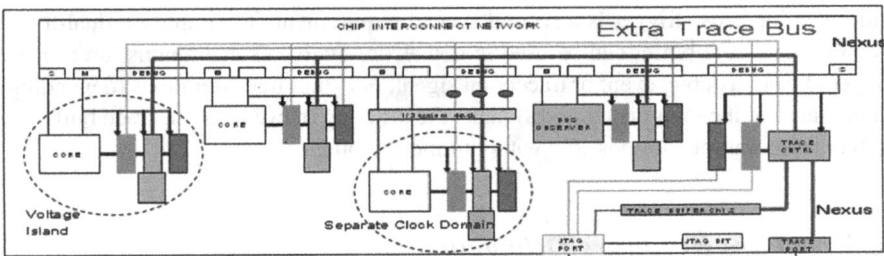

Fig. 10.3 Multicore synchronous debug implémentation

10.4 Debug Components and IP Interfaces

Figure 10.3 shows a simple system in which debug IP blocks that are socket inter-faces to various are integrated into a interconnect structure that is part of a bus fabric. All debug wiring goes through the system bus and is contacted through the OCP debug port. OCP debug ports may be implemented at points in the OCP sys-tem where a data port may not exist.

The programming of registers that contain either configuration or status informa-tion in the debug IP blocks may be JTAG-mapped or memory-mapped. Either or both modes of control and access are acceptable, based on specific system requirements.

In the memory-mapped case, the master port of the main debug core provides the programming of the debug block registers that have an address in the main memory space. The master OCP data port is not part of the OCP debug port. This permits one core to be accessed as the main debug agent of the system. The debugger sends instruc-tions via JTAG to this core agent and the core then accomplishes all actions in coordi-nating other cores through the main system bus. The core may be given special priority access within the system to unlock stuck interfaces and locked transactions and may initiate "Abort" and "Force" operations as part of the debug control interface.

In the more general system-independent JTAG-mapped register variant, the JTAG is part of the OCP debug interface. The debugger sends instructions to the cores over JTAG and the debug registers are part of a JTAG-TAP controller. Optional "Abort" and "Force" signals are also JTAG controlled.

For simplicity, discussion of the debug ports is limited to 1149.1 JTAG interfaces. The restriction is intended to simplify the port interfaces addressed. The intent is not to limit implementation to 1149.1 JTAG. Other bus interfaces, including those discussed in other chapters can provide similar access to the instrumentation.

10.5 Debug Socket Definitions

This section covers the basic signals and definitions for an OCP debug interface socket. An optional OCP port, known as the debug interface port, is added to all cores and IP blocks that support or need debugging access. The OCP debug port may be implemented as an addition to the OCP data (master and/or slave) port (in cases where debug blocks are memory-mapped, they may be controlled through the OCP data socket) or as an independent OCP port configuration.

The four basic groups of signals for a debug interface are:

- Debug control.
- JTAG.
- Debugger interface.
- Cross-triggering interface.

There are additional (extended) groups which can be optionally added to an instrumentation socket based on specific debug and analysis requirements. The optional extended debug signals in this interface are defined for optional debug features such as timestamps and performance analysis and to simplify definition of special "debugging-aware" functionality in designs that have security domains or power management with voltage islands.

Basic Socket-Level Debug Interfaces. Processor run control is typically implemented via the JTAG interface using debug mode signals in an IP. JTAG interfaces are supported in current OCP specifications with JTAG and related real time signals (Tables 10.1 and 10.2) for trace decoded at (one or more) JTAG TAPs (test access points).

A JTAG-only debug interface addresses many instrumentation operations that have non-real-time requirements by accessing debug-related registers on different cores. Even in the case of "memory-mapped" instrument blocks, the JTAG TAP and processor can typically have joint access and control of debug memory and registers for run control and monitoring operations. JTAG is also sufficient to configure and synchronize an OCP system into a trigger and trace modes. As a lower-speed serial

Table 10.1 OCP debug clock, reset interface socket

Debug interfaces	Description	Comment
Debug_reset_n	Debug clock source for instrumentation operation and optional debug system reset	Defined to be separate from system clock, reset so that debug can occur during system reset operations
Debug_en	General enable for debug modes	System input

Source: OCP IP

Table 10.2 OCP debug JTAG interface socket

JTAG Interfaces	Description	Comment
Tck, Trst (optional)	JTAG TCK, JTAG reset	JTAG input
TMS	JTAG TMS	JTAG input
TDI	TDI from previous node in JTAG loop	JTAG input
TDO	TDO to next node in JTAG loop	JTAG output
RTCK (optional)	Return clock signal for adaptive clocking	JTAG output

Source: OCP IP

interface, however, JTAG limits data-intensive debug operations such as trace, which is required in higher-performance test and debug interface efforts.

Higher-performance debug architectures may include independent reset and independent clock signals for the debug system synchronized to the debugger interface. An independent clock allows more flexible support of asynchronous or clock gated systems. An independent reset allows analysis of the target during system reset sequences. Additional reset and clock signals for timestamping counters may be common or independent from the debug control interface.

The use of JTAG debug interfaces is supported via the OCP 2.0 and higher specifications, and it is assumed that any JTAG signals are decoded at the core-level JTAG TAP. A JTAG-only debug interface does limit the ability to interface debug components on different cores and to set up and synchronize an OCP system into a debug mode.

Ideally, debug control signals are independent of the target system and have to duplicate many basic controls. The basic debug signals include an independent reset and independent clock signals for the debug system synchronized to the debugger interface. The reset and clock signals for timestamping counters are also part of this debug control interface. Debug clock, Instrument reset, and timestamp reset may also in many circumstances be common with external system clock or reset signals.

Even in the case of "memory-mapped" debug blocks, the processor control typically goes over this JTAG port.

10.5.1   Core Debug Socket Interfaces

These are connections to a processor or other OCP core to establish a debug connection. Core debug signals are generally specific to processor instrument interfaces and may be wrapped over existing debug signals for a core, except signals NoSResp and ForceResp, which may be mapped as extensions of OCP SResp signals. Because this socket is implemented for each bus master, different cores may have different control signals depending on the underlying functions that are supported by a debug block: for that core.

Table 10.3 defines a set of debug interfaces that address system-level debug of run control and debugger tool interfaces. Debugger access can therefore be consistently controlled via the debug interface signals. Not all signals may be required for all cores or systems:

- Special signals that support unlocking of stuck situations and forcing completion of locked actions (NoSResp, ForceResp, ForceAbort, ForceAbortAck) are here.
- Debugger accesses are qualified through MReqDebug.
- The processor acknowledges debug state entry through MSuspend.
- An OCP target can be configured to be sensitive to MSuspend.

Table 10.3 OCP debug interface socket

Minimum OCP debug signals set		
Signal name	Signal definition	Comment
MReqDebug	Qualifies an OCP request initiated by the debugger. MReqDebug may be a processor-native feature	If MReqDebug is derived from processor debug acknowledge, the OCP interface shall ensure there are no outstanding application transactions when the debug state is acknowledged.
Msuspend	The processor acknowledges the OCP initiator agent that is entering the debug state	The OCP initiator debug state acknowledge is routed to the OCP target. A debugger-aware peripheral may freeze a local HW process when the host enters the debug state
DebugSerror	Out-of-band error	Originated by debugger
DebugCon	The debugger is connected	Enables the on-chip debug hardware to communicate with debug host or agents
NoSResp	The target is not responding	If request has failed (ie. Sresp = FAIL), an indicates that the current transaction will not complete
ForceResp	The debugger has programmed the subsystem to force a data-independent response	No side effect to other threads
ForceAbort	The debugger has programmed the subsystem to solve the hang scenario	OCP interconnect handle abort without debugger intervention even in the case when the application SW has not enabled a time-out. The key requirement is to complete the transaction to allow the processor to enter the debug-state
ForceAbortAck	Acknowledge sent to subsystem (or debugger?)	Requires a Mabort input support in the OCP fabric to propagate the abort originated by the debugger to the initiator and OCP interconnect

Source: OCP IP

A debug component is informed that the debugger is connected through the DebugCon signal. A subsystem is informed that its TAP has been enabled by the application security software through the TAPenable signal. Depending on Debug-Mode[1:0], the debugger can initiate OCP transactions qualified as MReqSecure. (operations only enabled to qualified users as shown in Table 10.10).

Gated clock domains and voltage domains are used for power management in many ICs. A concern in system debug is to ensure that debug and trace operations are not interrupted or distorted by clocks or power to a connected system being removed at inappropriate times. OCP 3.0 defines a power management state machine for controlling connection and disconnection of IP blocks in preparation for power management, that allows main sections (as example, data transfer operations) and alternate sections (including debug operations) of a socket to be managed separately so that debug control signals in in case of gated clock domains and voltage domains are not interrupted if any IP block on the bus switches off clocks or voltages. By proper definition of the idling operations, such that a power down sequence does not occur when there is an active debug or trace operation in process, a blocking of the debug system shall not occur if one core or IP block goes into sleep mode.

Core (Master) Debug Socket Interfaces. OCP debug programming models should allow user-defined debug configurations based on the debug scenario and allow bidirectional debugger access to be consistently controlled via these debug interface signals.

Signals defined in the OCP-IP debug environment include debugger-initiated debug mode request (read/write) and core acknowledge signals to the debugger to communicate that a core is in the debug state. Because debug operations may interact with "normal" system operations, debug interfaces should also support unlocking of stuck-at situations and forcing completion of locked actions (force, abort, suspend) for a core in debug status. OCP peripheral interfaces would also need to be "debug aware" to recognize and synchronize with processor cores or other bus masters that enter debug mode.

Peripheral (Slave) Debug Socket interfaces. A peripheral debug interface should ensure that, for debugger OCP transactions, any debugger-initiated debug mode operation reads peripheral information transparently while preserving the system state. Depending on debug scenarios and the relationship between the local hardware process and the software process, peripherals should monitor the debug state and may need to take several actions to synchronize with the debug processor and to allow processing of OCP transactions initiated by the debugger to be handled differently than those initiated by the application software:

1. Freeze local hardware processes when the controlling OCP master is in the debug state. This may be accomplished by a parameter passed into a system debug register via JTAG or under software control, or it may be implemented as part of the debug hardware.
2. Stop peripheral or other local hardware processes when a processor enters the debug state. This can get complicated, because the peripheral may be shared or accessed by several OCP masters in the architecture.
3. Comprehend specific updating to ongoing local hardware processes when an OCP master enters a debug state; for example, disabling application-driven peripheral operations (such as flag clear, post-increment, and state machine updates).

To accommodate the diverse debug scenarios, a peripheral debug programming model may implement two or more debug control parameters in a system debug register as:

- FREE, which allows one/the program/etc. to keep the local hardware process running free and to make it sensitive to the debugger input.
- SOFT, which allows waiting for a clean boundary before stopping the local hardware process when extra latency is acceptable.

10.5.2 Cross-Triggering Socket Interfaces

Cross-triggering, and the associated system-level control, are important for debug of complex SoCs. Cross-triggering allows global and distributed event recognition and multicore triggering to identify and isolate events occurring throughout the system.

Information in multicore SoCs is complex and distributed such that global event cross-triggering and system-level control for multicore debug and triggering are often needed to identify and isolate events occurring throughout the system. Event recognition and triggering are widely used in conjunction with trace to capture information on events and operations in the SoC. Conditions are monitored and compared to generate real-time triggers in a cross-trigger manager. These triggers in turn can be used to control event actions such as configuration, breakpoints, and trace collection. More complex implementations can be programmed to trigger on specific values or sequences such as address regions and data read or write cycle types.

The cross-trigger block may be distributed to all IP connections to the OCP bus. If wiring is in the OCP fabric, then some pre-processing or wrappers (condition/action nodes) at each OCP interface can be used to simplify the cross-trigger information. Wrappers can be programmed via the JTAG debugger (or can be configured by a processor). Any block can send a trigger (edge or level) and receive a trigger. The debugger or processor can configure specific trigger lines for each IP to send a condition signal and from which trigger line it can receive a trigger/action.

Table 10.4 shows the OCP debug trigger interface socket. Each trigger line consists of two unidirectional signals and one (optional) enable signal. A minimum dual-channel concept consists of two independent trigger lines, but there is no upper limit on the number of cross-triggers realized in a design. The trigger line in, out or enable signals may be the result of a logic combination of several signals for a given core. Trigger lines may be connected directly to drive a bidirectional pin on the package and enable cross-triggering to continue between several chips. External (off-chip) triggers will be supported with pulse-width logic to interface external IO to the cross-trigger manager. Each debug channel needs one trigger line. The trigger logic grows linear with the number of cores or IP blocks that are debugged. No cross-trigger matrix is assumed necessary.

Event recognition and triggering is widely used in conjunction with trace to capture information on on-chip events and data in the SoC. Triggering conditions are monitored and compared to generate real-time triggers in a cross-trigger manager

Table 10.4 OCP debug trigger interface socket

Cross-trigger interfaces	Description	Comment
Trigger_in_ condition[m:0]	Trigger input from other OCP subsystems	X-trigger input shall support either high to low edge detection or level detection; during power-down of subsystem, trigger_in will not contribute to system cross-trigger action
Trigger_out_Action [n:0]	Trigger output to other OCP subsystems	X-trigger output of either active low pulse or level supports trace control or processor debug or interrupt request
Trigger_out_enable [n:0]	Optional trigger output enable to other OCP subsystems	Cross-trigger output
Ext_trig_clk	Optional Ext clock used for synchronizing trigger	External (off-chip or out-of-system) input
Ext_condition[n:0]	Optional Ext condition (e.g. debug status, tracepoint)	External (off-chip or out-of-system) input
Ext _action[n:0]	Optional Ext action (e.g. debug request)	External (off-chip or out-of-system) output

Source: OCP IP

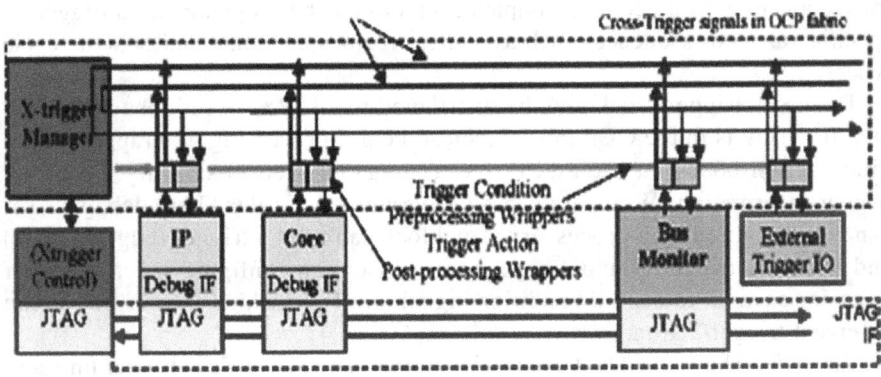

Fig. 10.4 Cross-triggering in the OCP bus

as shown in Fig. 10.4, where a cross-trigger with two "trigger lines" are looped back through the trigger manager.

Depending on the system configuration, signals may need to be preprocessed to allow conditions from different parts of the system to be synchronized or to support cross-triggering from external devices or external signals. Complex trigger implementations can be programmed to trigger on specific values or sequences, such as sequential combinations of rgw bus address region and data read or write cycle-type accesses.

Examples of triggering signals include debug or interrupt request conditions, although they can include any on-chip signal. Combinations of these triggers in

Table 10.5 OCP debug run control synchronization socket

Synchronous	Description	Comment
SyncRun	Synchronous run	Input
SyncRunAck	Synchronous run acknowledge	Output

Source: OCP IP

turn can be used to control on-chip actions such as core configuration changes, setting breakpoints or interrupts, initiating trace collection, or other user-defined requests.

The cross-trigger operation may be distributed among different IP connections to improve performance and support clock conversion and synchronization. Trigger-in (condition) and trigger-out (action) pre/postprocessing wrappers at each OCP interface point may be made synchronously configurable using signals in Table 10.5 to extend the cross-triggering abilities.

The triggering socket defines conditions, actions, and enables for on-chip trigger actions. Trigger operations may include processor-specific operations such as breakpoints and tracepoints, bus-specific operations such as trace sampling, and system-level interactions such as cross-triggers that may be applied to multiple cores, buses, and so on. There may be multiple instances of triggers (of varying complexity). The size of the condition and the actions are independent. Eeither selected trigger condition or trigged action may be either a single or a pattern of different lengths. Optional enables allow for selective condition monitoring (such as don't care situations) and global output actions:

Some general guidelines for OCP cross-triggering implementation are:

* Cross-triggering configuration shall be handled at the subsystem level.
* The subsystem can be programmed to:

  – Drive an OCP debug trigger-out line.
  – Be sensitive to an OCP debug trigger-in line.

* The OCP interconnect shall take care of the debug event triggers routing:

  – Point to point [1 trigger-out and trigger-in].
  – Broadcast [1 x trigger-out, n trigger-in].
  – Sharing [n trigger-out, 1 trigger-in].

* The OCP debug interconnect shall mimic a "tri-state" bus behavior through distributed combinatorial logic.
* An external device shall be able to contribute to cross-triggering.

The OCP cross-triggering configuration assumes that:

* Trigger-out (action) and Trigger-in (condition) routing for smaller implementations can be handled as sideband signals by the OCP interconnect.
* Trigger events may be routed to trace components (Table 10.6).
* Trigger events shall generate a user-defined request. This is typically classified as either a debug request or an interrupt request. These differ for different cores.

Table 10.6 OCP debug trace interface socket

Trace	Description	Comment
TraceTrigger[x]	OCP system event generates a trace trigger	

Source: OCP IP

- The OCP cross-triggering shall be operational for any platform subsystem frequency operating point supported by the cross-triggering configuration via level or pulse triggers.
- The OCP cross-triggering supports independent clock domains for trigger-out and trigger-in pulse conversion. Level triggering is recommended for widely varying clock domains.
- The OCP cross-triggering supports external trigger inputs from the IO pins. Triggers outputs can be routing to IO. Level or pulse triggers are supported with trigger pulse width modifiable to be compatible with device I/O performance.
- A subsystem in power-down or where debug has not been enabled shall be configured not to contribute to cross-triggering.

System observation using trace buffers and triggering on simultaneous events systemwide, including cross-triggering between chips, is a concept with limitations in time resolution that translate into distance limitations as described in the first approach. To overcome limitations in space we can give up precision in the feedback of the result as described in the second approach. To mimic a logic analyzer trigger, we need to have delay-equalized star-configuration to the trigger controller that will behave the same as in the second approach. Designers must decide which approach to take to create a consistent debug system Tables 10.4–10.10.

In cycle-exact trigger and feedback, it is crucial that collection of all trigger conditions complete synchronously to the highest system clock cycle. The advantage is that sequencing of trigger conditions that are one cycle apart is possible even at the trigger sources. The difficulty is to close timing in such a design because the trigger path becomes the biggest bottleneck on the critical timing path.

The trigger logic in the OCP debug socket is based on a distributed model of a tri-state wire. The trigger events are collected with a chain of distributed AND-gates and the result is sent back over a second wire in a half-loop arrangement. The trigger controller connects to "the last OCP debug socket" at the end of the AND-gates and loops the result back to the second wire.

Cycle-exact triggering accepts the feedback signal on the second wire to arrive in a later cycle to help with timing closure. This means that detection of a trigger equation has to occur over one cycle, but propagation of the trigger's action, for example to stop a trace buffer, can extend over several cycles. A delayed trigger action that is used to make event decisions requires that consecutive events that are several cycles apart to insure that the actions are fully propagated before the next event trigger occurs.

Aligning debug information in the display to be cycle-exact uses a local time-stamp during collection of trace information. Stopping the trace buffer a few cycles after a trigger condition will still allow for exact time alignment in the display.

The trick is to equalize the arrival time at the trigger controller from any trigger source by inserting delay buffers before entering the AND-gate trigger line. It is then possible to trigger on events that occurred at the same time. Sequencing of triggers that occurred one cycle apart is possible inside the trigger controller by using multiple arrival-time-equalized trigger lines. As a logic-analyzercan trigger on the acquired signals but does not supply trigger information back to the device under test, the OCP debug system with a relaxed feedback concept does not demand to have delay-equalized feedback connections back to the trigger sources. For triggers coming from all corners of a widely distributed on- chip trigger network or for systems where cross-triggering is required between chips, this provides a timing tolerant solution. It scales well to a system of any size and can have extra built-in arrival-delay of "several clock cycles" to accommodate triggers coming over external pins. The proposed OCP debug cross-trigger concept can be used for this configuration. The fixed built-in target trigger arrival delay is independent of the highest clock in one chip or in multiple chips.

Exact triggering in a star configuration is similar to a logic analyzer; the cycle-accurate trigger timing can be designed by delay-equalized trigger lines going to the trigger controller in a star configuration. This requires a separate trigger line from each possible trigger source. Any sequence of trigger events can then be realized as cycle-accurate inside the trigger controller. However, the feedback to the trigger sources, or to the assertion blocks, allowing them to perform cycle-accurate trigger sequencing remains an issue. Stopping pf trace may still occur a few clock cycles later. This star topology concept can be made cycle-accurate in any system at the expense of individual trigger lines with delay equalization. Clearly, this concept does not scale with large systems because wires grow proportionally to sources and not proportionally to trigger decisions. Star configuration is not part of this proposal because the arrival time equalization with the proposed distributed AND-gate trigger line will work equally well.

10.5.3   OCP Synchronized Run Control

Synchronized run control allows clock-synchronized program execution of two cores that would usually run asynchronously. This makes it possible to time-align the instruction streams to study interdependence.

When we debug several heterogeneous cores with different clock speeds, a single step needs a new definition. Stepping is no longer a single core operation, but also a problem of how to stop cores synchronously to events that are caused by a single core (for instance on a breakpoint hit). The debugger reaction depends on the core interaction scheme; for example, cores that are virtualized using SMP should be stopped synchronously by hardware within a few clock cycles. This is not a problem, because SMP cores are driven by the same clock domain. In isolated/loosely coupled multicore environments, the core's stop-timing is usually less critical, thus achieving the required synchronization latency through separately issued TAP commands. Hardware synchronization would be advantageous in case of higher latency requirements.

Hardly any multicore architecture on the market implements Single-Stepping by means of fetching and executing exactly one instruction on every core. Many heterogeneous architectures have cross-core Single-Stepping hardware implementation:

1. The system is halted. The debugger reads/has the full state of all cores/ memories.
2. Run about 100 to 1,000 cycles and halt synchronously.

Trace IP and all data accesses during this timeframe. (There is no problem with MCDS, even with a small trace memory.)

3. With this information, the debugger can exactly reconstruct all states and data values between the start and endpoint.

This allows the cores to virtually swap single-step operations with regards to each other in this time window. The timing relationship between the cores is well maintained. There may be only a slight impact at the start and end of the period.

10.5.4   OCP Traffic-Monitoring and Trace Interfaces

Traffic monitoring and trace are often critical debug features to be able to analyze on-chip behavior. System monitoring and trace can be performed at signals on the data socket or in the bus fabric itself.

Trace requirements are application-dependent, requiring signals and monitoring bus traffic events that may be extracted from the system cross-trigger information or provided by a processor or other on-chip IP. Trace should be noninvasive (should not affect OCP system behavior) and should be secure (should not allow unauthorized accesses into the instrumentation system). Useful features for bus monitoring and trace include:

1. Continuous (or at least long-duration) system monitoring.
2. Filtering based on OCP operations (e.g. initiator, thread, address range, DMA logical channel).
3. Trace capture of both OCP transactions and non-OCP qualifying events.
4. Transaction filtering and alignment of requests and responses.
5. Elastic trace bandwidth at OCP system traffic peaks.
6. Support for SW instrumentation interleaving with the trace flow.
7. Support interleaving several trace flows from different trace points or channels.
8. Support multithreaded data observation, including system trace data reads from the JTAG or from application SW. Because trace is data-intensive, high-performance interfaces may be required.

Trace triggers provide trace enable and control for OCP bus and logic trace and as a performance and analysis interface to specific internal event. In the case of real-time tracing to outside pins, specific trigger signals are included in the trace and/or performance monitoring (Tables 10.6 and 10.7) interfaces of the OCP debug interface socket. The trace trigger is extracted from the information on the cross-trigger lines.

Trace-packet interfaces are defined in several protocols, including Nexus (IEEE 5001) and MIPI. Because there are other standards bodies addressing these issues of higher-performance debug interfaces, OCP debug leaves this level of interface open to the user's preference.

Traffic Monitoring and Trace – General Configuration:
The OCP system monitoring debug instrument allows monitoring of the "OCP system" bus traffic:

- Focus on specific OCP transactions.
- User-defined transaction filtering.
- Initiator, thread, address range, DMA logical channel.

An emulator or debugger host to configures the OCP system monitoring component from the external [JTAG] interface through the OCP debug bus.
The OCP system monitoring instrumentation allows:

- Alignment of the OCP transaction requests and responses.
- Capture of the additional OCP transaction qualifiers.
- Export of the captured traffic data through the OCP debug to a trace export component.
- Support of continuous system monitoring.
- Preservation of the OCP system bus behavior.
- Have options for securing the systems from unauthorized accesses.

Table 10.7 OCP debug performance-monitoring interface socket

Performance monitoring	Description	Comment
MConnID	Identifies the initiator. Routed to target	Determines active initiator for monitoring
MChannelID	Identifies the DMA channel initiator. Routed to target	Determines active channel for monitoring
MReqWatch[x]	Qualifies an OCP request	
PMSampling	Periodic performance metric sampling	Initiates a periodic transfer of the performance metrics computed by a system interconnect instrument to atrace export component. Operations assume that the Periodic sampling strobe is generated within the OCP instrument

Source: OCP IP

The trace export interface may contain a variety of different features:

- Implement an elastic buffer.
- Optionally build trace packets for different (MIPI/Nexus) protocols.
- Support a trace export bandwidth compatible with OCP system traffic peaks.
- Allow SW instrumentation interleaving.

The trace buffer instrumentation may support several modes of operation:

- Provide flexibility to disable capture around a trigger.
- Allow system trace data reads:

  - From the JTAG-OCP component.
  - From the application SW.

- Allow interleaving several trace flows.
- Allow multithreaded data observation.

10.5.5 Performance Monitoring

Performance monitoring enables observation of selected threads, initiators, and targets to identify data traffic and measure data bandwidth.

OCP performance monitoring requirements vary widely and are by nature application-specific. Some genral signal examples are given in Table 10.7, for sample based performance monitoring. Following is a general set of requirements for a performance monitor that supports many common analysis requirements:

- An OCP debug component allows monitoring of the OCP system bus bandwidth.
- An emulator shall be able to configure the OCP performance monitoring component from the external [JTAG] interface through the OCP debug bus.
- OCP initiator transactions monitored for different OCP targets.
- Monitor task windows [start and stop triggers].
- Monitor system event latency between two selected signals (using timestamp or other counter logic).

The OCP performance monitoring instrumentation many be used in several ways:

- Count within the [start, stop] window defined by triggers:

  - Effective cycles at the OCP target level.
  - Waiting cycles at the OCP initiator level [latency, arbitration, shared link, etc.].
  - Free cycles at the OCP target level.

- Support continuous performance monitoring [statistics].
- Export the computed performance statistics data through the OCP debug to the trace export component.
- Preserve the OCP system bus behavior.

The trace export instrumentation may include feature to:

- Implement an elastic buffer.
- Optionally build trace packets for different (MIPI/Nexus) protocols.
- Allow SW instrumentation interleaving.

10.5.6 System Timestamping

For distributed systems, a timestamp provides the means of temporally correlating different events that may be occurring in different systems or domains. There are many timestamp implementations – the simplest is a gated clock and reset that can be used to run timestamp counters at different blocks, which is shown in Table 10.8. This interfaces assumes that other parameters (timestamp length, mode, etc.) are hard coded or pre-defined elsewhere.

Synchronous start of local timestamp counters is required for accurate distributed local timestamping. The synchronized start of all local timestamp counters is important for the correct display of debug events. The frustration on debugging a timing problems that turns out to be an artifact of trace synchronization cannot be overstated. Two basic rules should be followed where possible:

(a) If operations start on reset being released, use an asynchronous reset, where the releasing edge is synchronized to the slowest clock or at times when clocks coincide with their rising edge, so that it arrives at the same time to all registers, regardless of the local clocks speeds. If all local clocks are time-aligned and iso-synchronous of each other, then this will ensure that all counters and other start logic are aligned. Obviously this requires knowledge of the local clock frequencies and their skew at the time of reset-release.
(b) It is always best if one balanced clock goes to all counters at the same time and is supplied only while tracing is active. Ideally, this one clock issynchronous to or multiple of all local clocks. Otherwise, it requires a fair amount of over-clocking to resolve phase relationships between the asynchronous clocks.

Stamp clock and stamp reset signals are both part of the basic OCP debug interface, but should be implemented with care to provide synchronous capture of debug data.

Table 10.8 OCP debug timestamp interface socket

Timestamp interfaces	Description	Comment
Ts_enable	Timestamp start and stop	May be driven by trigger logic
ts_clk	Timestamp clock (gated version of clk) for global on-chip timestamping	Timestamp clocks do not necessarily need to be the system clk
ts_reset	Timestamp reset	Should be different from system reset

Source: OCP IP

10.5.7 Power Management Monitoring

Power management by reducing or turning off the clock and switching off the power supply to certain IP blocks is increasingly required in many systems. It is important that debug operations not get locked or interrupted while dealing with power-aware IP blocks transitioning their power state during a debug session. Power management Debug signals (Table 10.9) monitor the power state of each socket in the system. Since power states may be different for primary and auxiliary logic in a given block, the OCP Sresp signal is extended to include no power and no clock output states for cases where the debug interface or other portions of the target are in power down or not receiving a clock.

OCP-IP in its 3.0 bus socket architecture release has defined four state FSM-based power-down sequences for each master. Because each master may be powering down and up on its own schedule, signaling from a given core for system debug operations can be very dynamic. The GFSM states allow auxiliary sockets (of which debug is one) to be powered down in a quasi-independent means from the main socket interface. Therefore it is typically possible to have a debug socket remain active even if the core to which it is attached is powered down.

The OCP platform power management module generates a trigger when:

- Switching off a domain.
- Waking up a domain.
- Switching frequency.
- Switching operating voltage.

The OCP power management monitoring may:

- Support continuous power management monitoring.
- Preserve the OCP system bus behavior.
- Not require SW instrumentation.

Table 10.9 OCP debug power monitoring interface socket

Power management	Description	Comment
Sresp[2:0]	Additional error response codes signal a target is not powered or not clocked	NULL, DVA, FAIL, ERR – new codes NOCLK, NOPWR
PWRDomainStatus	Indicates to target agent if power domain is active	PWr status signals contribute to error response generation
CLKDomainStatus	Indicates to target agent if clock domain is active	Clk status signals contribute to error response generation

Source: OCP IP

10.5.8 Security Debug Interface

Security concepts require enabling debug of sensitive locations only during authorized chip access and disabling it otherwise. Debug signals, with their access to a
wide range of system data, need to be secured. OCP security signaling is here
extended to the debug socket so debug IP blocks can implement a general lock-
down unless there is qualified access. Because the specific security methods vary
widely, the interface methods are generic, with Table 10.10 providing a OCP
debug security interface socket that allows general control and status for secure
mode operations.

OCP instrumentation framework concepts are extensible to multi-channel event
synchronous debug and can be applied to a range of situations, from single-core
debug to large numbers of core and even subsystems. The general instrumentation
architecture and sockets can be extended to more debug channels by duplicating
hardware.

Table 10.10 OCP debug security interface socket

Security	Description	Comment
MReqSecure	Qualifies an OCP request as a secure transaction	The application security setup [HW and SW] may allow qualifying debugger access as secure
DebugMode[1:0]	Debug operating mode	Debug can be disabled, restricted to public OCP transactions, or allowed for both public and secure transactions
TraceMode[1:0]	Trace operating mode	Trace can be disabled, restricted to public OCP transactions, or allowed for both public and secure transactions
TAPenable	Subsystem test access port	Enabled by application security software

Source: OCP IP

Chapter 11
Nexus IEEE 5001

Nexus 5001 is a debug standards initiative based on the IEEE ISTO 5001 debug specification that addresses the diverse challenges for embedded-processor and digital-system debug interfaces. To address applications (data communication, automotive powertrain, computer peripherals, wireless systems, and other control applications) with constantly increasing complexities requires more comprehensive debug features and will benefit from more standardized interfaces. As advances in semiconductor and system design continue, these types of embedded applications use higher-performance embedded processors. Nexus 5001 was defined in 1999, and its development and proliferation are managed by the Nexus 5001 Forum™, which evolved as a successor to the Global Embedded Processor Debug Interface Standard Consortium (GEPDISC), which was formed to develop an embedded-processor debug interface standard for embedded control applications. The latest version of the Nexus standard was released in 2003, with ongoing work aimed at a new release in 2011. Nexus architectures have been used extensively in U.S. automotive applications, and more chips have been produced incorporating Nexus ports than any other nonproprietary debug-specific interface. The Nexus 5001 Forum is an industry-based standards group that manages the IEEE 5001 (Nexus) debug specification.

Efficient use of these embedded processors requires software and hardware development tools that can easily access critical processor functionality. The lack of a unifying standard among the various embedded processors on the market has impeded this accessibility, preventing tool vendors from creating standard tools with consistent functionality across a broad range of processors. Nexus 5001 addresses this issue by providing a consistent set of auxiliary pin functions, message-based transfer protocols, and standard development features as shown in Fig. 11.1 to facilitate debug implement. The standard itself is open and processor-independent, but the implementations are user-specific. The full release of the Nexus 5001 specification is freely available for download from the Nexus Web site at http://www.nexus5001.org/.

The Nexus architecture defines high-performance data interface, protocol, and register infrastructures that can be used to implement a variety of trace and control instrumentation (Fig. 11.2). The Nexus infrastructure includes features that support multicore development and multifeatured trace and configuration/control.

N. Stollon, *On-Chip Instrumentation: Design and Debug for Systems on Chip*,
DOI 10.1007/978-1-4419-7563-8_11, © Springer Science+Business Media, LLC 2011

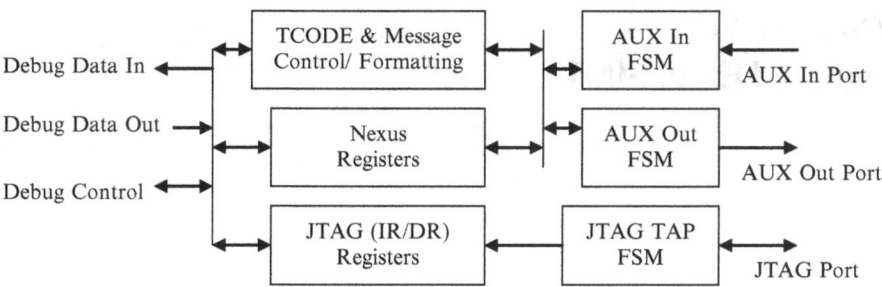

Fig. 11.1 Nexus internal architecture

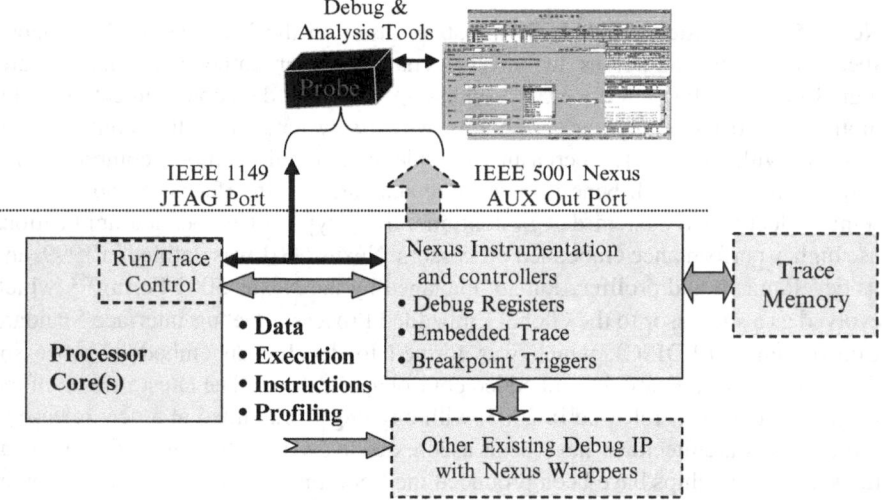

Fig. 11.2 Nexus interfaces

Nexus architecture is based on a packet-based messaging scheme, which supports debugging complex multicore systems. Control of the multicore debug processes based on a transaction protocol (TCODE) that allows data to be sent in packets, using a packet header to provide information on the source and assumed destination of the data on-chip components as well as information on the subsequent data packets containing trace or other information. This simplifies interleaving of multiple trace sources and concurrent communication with multiple Nexus instruments. The Nexus specification defines a standard set of TCODEs for common identification and trace operations; the TCODE protocol is also extensible to user-defined debug commands (see Table 11.4).

Nexus also defines a standard set of debug-related on-chip registers, which facilitate the identification, communications, and interfacing to different cores and subsystems for multicore control and debug operations. A standard register set allows simpler integration and control of the instrumentation with embedded debuggers and related tools.

The Nexus 5001 Forum is engaged in ongoing collaboration with other industry debug-related efforts, including OCP-IP and MIPI, and is in the process of extending the IEEE 5001 specification to support emerging debug interfaces such as SERDES and two-wire JTAG (1149.7) ports to address diverse debug requirements.

11.1 Nexus Implementation Classes

Applications have varying debug requirements, but most debug can be grouped into performing certain classes of tasks. Nexus defines debugger functionality and compatibility over four classes of operation. Device instrumentation and tools are defined as being class 1– to 4–compliant if they support all of the features defined for that class. Class 1 starts with basic debug functions over a JTAG port, with higher classes involving more instrument access and system complexity using the AUX port to progressively increase debug capabilities, such as adding more complex trace and emulation analysis of processor operations.

Features in the Nexus implementation classes can be customized so that designers can select features of importance and not be burdened with more advanced features or those that are not applicable or efficient to their debug needs. This allows a variety of debug features to be supported, while keeping the number and types of different Nexus implementations that need to be tracked and supported manageable. All Nexus classes by definition include all of the features in (i.e. are a superset of) the prior class(es). The key features of the different implementation classes are summarized in the Table 11.1.

The most basic, class 1, provides features similar to standard JTAG implementation. Class 1 provides run-control debug features that are common with most processor implementations, including core identification, single stepping,

Table 11.1 Nexus 5001 implementation classes

Nexus	Services	Features
Class 1	Static debugging	Single step
Basic run control	Breakpoints	Set breakpoints and watchpoints
		Two breakpoints minimum
		Device identification
		Static memory and I/O access
Class 2	Watchpoints	All class 1 features
Instruction trace	Ownership trace	Monitor process ownership in real time
Watchpoints	Program trace	real-time program tracing
Class 3	Data trace	All class 2 features
Data trace	Real-time read/write	Access memory and I/O in real time
Read/write access	Transfers	Real-time data tracing
Class 4	Memory substitution	All class 3 features
Memory and port substitution	Port replacement	Start traces on watchpoint occurrence
		Program execution from Nexus port

Source: Nexus 5001 Forum.

breakpoints and watchpoints, and static memory and I/O access. Class 1 has certain minimum requirements, such as the need for at least two hardware breakpoints. Debugging halts the chip while commands are executed.

Class 2 contains more complex debugging features with real-time monitoring. It also adds instruction tracing and more sophisticated watchpoints. Class 2 enables processor execution trace-related features including real-time monitoring of process ownership and instruction tracing, along with complex watchpoints and branch tracking, flagging indirect branches, and eliminating redundant addressing information. The class 2 program trace feature allows indirect branches to be flagged, making it easier to differentiate indirect branches from exception-handling operations. Additional messages are included for improved branch tracking. The format of the trace data allows for the elimination of redundant addressing information, which increases throughput.

Class 3 allows data-tracing services and includes the ability to read and write memory and I/O while the processor is running. Class 3 supports data tracing and memory and I/O read and write while the processor is running. This makes the system design more complex, but significantly improves the debugging capabilities.

Finally, class 4 delivers features found in many in-circuit emulators (ICEs). Class 4 allows direct user control of a processor to execute programs from the Nexus port (memory substitution), plus additional features for remapping memory and I/O ports and starting trace on watchpoint occurrence. This is especially useful when simulating peripherals. It can also be used to provide other applications running on the testing system with access to shared memory. Class 4 features include starting memory substitution on watchpoint occurrence, monitoring data reads while the processor is running in real time, port replacement and port sharing, and the ability to transmit data values for acquisition.

11.2   Nexus Message Architecture

Nexus messages consist of a 6-bit TCODE that contains Nexus-specific instructions followed by a variable number of packets (the number of packets for each TCODE is defined in the standard). Messages can be sync or nonsync. Sync messages include the full address and nonsync only include relative address changes. Each message also contains a SRC field (source ID) to help development tools identify the source of a particular Nexus message in a multiprocessing SoC sharing a single debug port. Packet types supported include the following:

Variable: A variable-size packet means the message must contain the packet but the packet's size may vary from a minimum of 1 bit. An example is an address field that may be full or partial for a given message. When messages are transferred via the AUX, variable-size packets must end on a port boundary.

Vendor-fixed: These are used to allow Nexus packets in to match characteristics of a vendor's device. An example is a SRC field that identifies the source ID;

vendor-fixed packets may be of zero length (not implemented if not required, as in the case of a Nexus system with only one on-chip instrument, where the message source can be assumed).

Vendor-variable: These are used to allow Nexus packets to match characteristics of a vendor's device. Vendor-variable packets may be of zero length (i.e. not implemented). An example of a vendor-variable field is message timestamp. When messages are transferred via the AUX, vendor-variable packets must end on a port boundary. Variable-size packets may have different lengths in messages of the same type, so MSE signaling protocols are used to determine the end of packet boundaries. Typically vendor-variable packets are target-processor-dependent and have a variable size determined by the processor vendor. These packets are normally reserved for the end of a public message where the vendor may implement additional fields.

All Nexus TCODES follow a common message format. An example of a Nexus message, program trace with indirect branch shown in Table 11.2, consists of the TCODE = 4 followed by a message-specific number of packets of differing types. Complete descriptions of all the message types and their options are given in the Nexus specification.

TCODES can be either public (defined in the Nexus standard) or user-defined. Public TCODES defined in the Nexus standard (IEEE-ISTO 5001-2003) include a range of trace options as well as other Nexus operations. Only a subset of the total available messages must be implemented in a given system. The minimum required messages for an implementation are given in Table 11.3.

Table 11.2 Example packet fields in trace message

Program trace – indirect branch message			Direction: from target
Minimum packet size (bits)	Packet name	Packet type	Description
0	TSTAMP	Vendor-variable	Number of cycles message was held in the buffer or the full timestamp value. For targets that do not implement timestamping (or use pins for timestamping), this field may be omitted. Refer to 4.11.2 – Timestamping via AUX
1	U-ADDR	Variable	The unique portion of the branch target address for a taken indirect branch or exception
1	I-CNT	Variable	Number of instruction units executed since the last taken branch
0	B-TYPE	Vendor-fixed	Branch type. For targets that do not need to differentiate branch types, this packet can be omitted
0	SRC	Vendor-fixed	Client that is source of message. For targets with only a single client, this packet can be omitted
6	TCODE	Fixed	Value = 4

Source: Nexus 5001 Forum.

Table 11.3 Minimum required public messages

Message type	Compliance class	Minimum required public messages
Device ID	2, 3, 4	Device ID
Ownership trace	2, 3, 4	Ownership trace
Program trace	2, 3, 4	Direct branch, indirect branch, synchronization[1], error
Data trace	3, 4	Data write, data write message with sync, error
Read/write access	3, 4	(1) For embedded processors that implement the recommended registers: Target Ready, Read/Write Register, Read/Write Response
		(2) For embedded processors that implement device-specific registers: read target, write target, read next target data, write next target data, target response
Watchpoint	2, 3, 4	Watchpoint message
Memory substitution	4	Read tool, read next tool data, tool response

11.2.1 Nexus TCODEs

Nexus TCODEs can be classified into six different types, which are described in detail in the Nexus specification. Table 11.4 provides a summary of the packet fields that are used for different TCODES. Different TCODE classes include the following:

1. Status indicates status information messages from the target. This group includes register reads and core-specific or watchpoint/breakpoint status, error messages, and so on (TCODEs 0–2, 8, 15).
2. General register read/write is a group of commands that allow memory-mapped reads and writes between tools and Nexus recommended registers (NRR) or other registers in a Nexus-defined memory map. Among other general applications, these messages can be used for run control and configuring watchpoint/breakpoint operations (TCODEs 16–19).
3. Program trace is a range of trace options that rely on Nexus-defined branch trace schemas, which limit instruction trace to discontinuities (branches, conditional jumps, interrupts, etc.) and their relative distance from the last trace. By mapping these values to an assembled program, debuggers can interpolate branch locations in the program flow and reconstruct (interbranch) instruction flow. Nexus also defines periodic sync fields and trace messages to identify inconsistencies and align trace, which is useful in correlating execution over multiple cores (TCODEs 3, 4, 9–12, 27–33).

   • Program trace:

     – Direct branch.
     – Indirect branch.
     – Indirect branch with history.
     – Synchronization.
     – Resource full.

Table 11.4 Nexus messages and parameters

Nexus command type	Nexus message name	TCODE [5:0]	Direction	TSTAMP	STATUS	SRC	ID	INST	ADDR	TYPE	BTM	DATA
Status	Debug status	0	From target	y	DS reg	y						
	Device ID	1	From target				DID					
	Ownership trace	2	From target	y		y	TASK ID					
Basic program trace	Program trace – direct branch	3	From target	y		y		I_CNT				
	Program trace – indirect branch	4	From target	y		y		I_CNT	U-ARRR	B-TYPE		
Data export	Data trace – data write	5	From target	Y	DSZ	y		DCORR	U-ADDR	MAP		DATA
	Data trace – data read	6	From target	Y	DSZ	y		DCORR	U-ADDR	MAP		DATA
	Data acquisition	7	From target	y			IDTAG					DODATA
	Error	8	From target	y		y	ECODE					
Synchronized	Program trace – synchronize	9	From target	y		y	DCONT	I_CNT				PC
Program trace	Program trace – correction	10	From target	y		y		ADJUST				
	Program trace – direct branch with sync	11	From target	y		y	DCONT	I_CNT	F-ADDR		CANCEL	
	Program trace – indirect branch with sync	12	From target	y		y	DCONT	I_CNT	F-ADDR	B-TYPE	CANCEL	

(continued)

Table 11.4 (continued)

Nexus command type	Nexus message name	TCODE [5:0]	Direction	TSTAMP	STATUS	SRC	ID	INST	ADDR	TYPE	BTM	DATA
Data trace	Data trace – data write w/sync	13	From target	y	DSZ	y		DCORR	F-ADDR	MAP	CANCEL	DATA
	Data trace – data read w/sync	14	From target	y	DSZ	y		DCORR	F-ADDR	MAP	CANCEL	DATA
Watchpoint access	Watchpoint match	15	From target	y	WPHIT	y						
Register access	NRR access – target ready	16	Both ways	NO FIELDS								
	NRR access – read register	17	From tool			OPCODE						REG VAL
	NRR access – write register	18	From tool			OPCODE						REG VAL
	NRR access – read/ write response	19	Both ways									
Port replacement	Port replacement – output	20	From target		DIR							OUT DATA
	Port replacement – input	21	From tool									IN DATA
Memory access	Memory access – read target/tool	22	Both ways		DSZ				ADDRESS	MAP		
	Memory access – write target/tool	23	Both ways		DSZ				ADDRESS	MAP		DATA
	Memory access – rd next target/tool data	24	Both ways	NO FIELDS								
	Memory access – WR next target/ tool data	25	Both ways									DATA
	Memorty access – target/tool response	26	Both ways		ST							DATA

Category	Message	TCODE	Direction		Packet fields					
Advanced trace	Program trace – resource full	27	From target	y	RCODE					RDATA
	Program trace – indirect branch history	28	From target	y	I_CNT	U-ADDR	B-TYPE			HIST
	Program trace – indirect branch history wiSync	29	From target	y	DCONT	I_CNT	F-ADDR	B-TYPE	CANCEL	HIST
	Program trace – repeat branch	30	From target	y	B-CNT					HIST
	Program trace – repeat instr	31	From target	y	I_CNT	F-ADDR			R-CNT	HIST
	Program trace – repeat instruction with sync	32	From target	y	I_CNT	F-ADDR			R-CNT	HIST
	Program trace – correlation	33	From target	y	EVCODE	I_CNT				CDATA
Undefined by spec	Reserved	34–55	Both ways							
	Vendor-defined message	56–62	Both ways							
	Vendor-defined extension msg	63	Both ways							

- Repeat branch.
- Repeat instruction.
- Correlation.

4. Data trace is a trace of data values associated with a defined address range for efficiency. Nexus also supports data-acquisition instructions for streaming export of larger amounts of system information; such as data from on-chip buffers or FIFOs (TCODEs 5, 6, 7, 13, 14).

- Data trace:

  - Data write.
  - Data read.

5. Memory access is a nonintrusive peek and poke operation of internal memory blocks; it can also be used for directly driving from a Nexus memory or location (TCODEs 22–6).
6. Port replacement allows Nexus pins to emulate other I/O functions of comparable speed (TCODEs 20, 21).

User-defined TCODEs can be defined by silicon or IP developers to add additional debug features not covered by the standard, similarly to user-defined instruction features in JTAG.

11.2.2   Nexus Registers

Nexus also defines a standard set of debug-related on-chip registers, which facilitate the identification and interface to different cores and sub-systems and to multicore control and debug operations. A standard register set allows simpler integration and control of the instrumentations with embedded debuggers and related tools.

Nexus defines a number of recommended registers, which facilitate the integration of debug support to different cores. Of particular interest for multicore designs, each core or element on a device may be assigned a different ID in a device identification (DID) register to allow discrimination and selection of control and debug operations associated with a given block or subsystem.

Nexus defines and assigns register maps to 63 recommended registers, which are accessed by TCODE operations. Different instances of the same register can be associated with different cores by a source field value that can be transmitted as part of each output message. NRRs may contain recommended fields, specifying control or status information, and may include the following:

1. Device identification registers are IDs for discrimination and selection of different sub-systems (at the SoC level) or at the chips (for multichip debug scenarios). This register provides device configuration information similar to what is provided for 1149.1 JTAG DID access, which is a required JTAG instruction.
2. Client-select register (CSC) contains information on the originating source (i.e. processor or core) for trace and other exported messages.

3. Control register (DC) contains debug parameter and configuration information.
4. Status register (DS) contains debug status information.
5. User base address register (UBA) defines the base address for relative or truncated addressing modes.
6. Watchpoint trigger registers (WT) provide watchpoint or breakpoint status.
7. Data trace attribute registers (DTSA/DTEA/DTC) contain information on recent trace operations and program information needed to reconstruct the trace.
8. Breakpoint/watchpoint control registers (BWC) contain watchpoint and breakpoint configuration information.
9. Breakpoint/watchpoint address/data registers (BWA/BWD) define address and/or data for assigning watchpoint and breakpoint locations.
10. Read/write access registers (RWA/RWD/RWCS) contain the information used for memory-access operations.

Optionally, the two BWC registers may be combined with the two data trace attribute registers so that a total of two registers may be simultaneously active; that is, two BWC registers, two data trace attribute registers, or one BWC register and one data trace attribute register.

Most processor debug environments can be adapted to be Nexus-compliant by adding a Nexus wrapper layer around the existing debug blocks. The value of Nexus for processor debug is that it allows a consistent environment for different processor types to be integrated using a consistent methodology.

Nexus defines a method of trace compression that takes advantage of the properties relating to execution of instructions being pre-defined during the programming; unlike many other types of trace operation, it is largely deterministic. With the exceptions of branching and other instructions that are conditional on data, the sequence of instructions through a processor is pre-defined during software development.

To make efficient use of memory resources during execution trace, Nexus uses a processor instruction compression technique called branch trace messaging, which reduces the trace memory required by focusing, capturing only a full trace on instruction flow discontinuities (typically branches). Because branches and conditional operations typically constitute a small percentage of an overall instruction execution, this can greatly expand the trace RAM utilization. There are other conditions from which trace information can be tightly integrated with debugger software tool chains to allow correlate analysis of the source code. Nexus also supports relative addressing to reduce the number of required address bits transmitted for normal messages. Certain initialization and exception cases (defined within the standard) will cause normal trace messages to be "upgraded" to sync-type messages in which the entire address is transmitted. Execution trace can be compressed and later expanded for integration with code debugger tools. This feature allows debug blocks storing instruction trace to leverage assumptions in instruction flow in order to conserve trace bandwidth and increase the number of instructions that can be stored in trace buffers or exported in real time.

For data trace operations, other than the use of relative address transmission (as in program trace), there is typically no such determinism that can be leveraged for the data itself to extend the use of trace resources, and as such data trace may require either larger trace memories for a given trace size or alternative methods of storing trace information.

Even with compression, the time needed for trace export can be significant when relying only on JTAG TDO to transmit data. This problem increases proportionally for multicore designs, where each processor and other block has its own debug information. Improving trace interface throughput is a primary reason for implementing a Nexus AUX port.

11.3 NEXUS Interfaces

Nexus provides a standardized interface for on-silicon instrumentation and debug tools providing a range of expanded features for system debug. Most notable are higher-performance auxiliary interfaces to support real-time and data-transfer-intensive operations such as trace.

At its simplest level, Nexus is compatible with JTAG but recognizes that the limitations in JTAG bandwidth are not realistic for the debug requirements for complex or multicore environments, and provides options for both input and output auxiliary parallel interfaces for high-speed data transfers. The Nexus specification defines a vendor-neutral IO signal interface and communication protocol that supports parallel debug and instrumentation support. The Nexus interface defines a small set of control signals and AUX data ports (shown in Table 11.5) that are implemented in conjunction with JTAG or as a self-contained port. The additional data pins provided by the AUX interfaces are scalable for matching the debug requirement and allow much higher read/write throughput between the target and debug and analysis tools compared to JTAG Figs. 11.1 and 11.2.

11.3.1 Nexus JTAG Access

Nexus messages may be read from or written via the IEEE 1149.1 JTAG port. Message writes are generated by an external IEEE 1149.1 controller and are input into an input public message register (IPMR). The IPMR receives its TCODEs and packets via multiple passes through the SELECT-DR_SCAN.

The IEEE 1149.1 protocol does not permit public messages to be initiated from an on-chip interface. Therefore, an output public message register (OPMR) is available for transmission of messages from the embedded target microcontroller to an external IEEE 1149.1 controller.

The JTAG port is used in Nexus-specific ways to implement various classes of services such as reading and writing Nexus registers and messages, and allowing Nexus trace output to be embedded into JTAG messages. Output public message reads are messages that are generated by the target processor and are read from the OPMR. These unsolicited messages may contain variable-length packets of data. Two methods may be used for determining when an output public message is

Table 11.5 AUX interface signals

AUX IO	Description of auxiliary pins
MCKO	Message Clock out (MCKO) is a free-running output clock to tools for timing MDO and MSEO pin functions. MCKO can be independent of the embedded processor's system clock, or an embedded processor's clock pin may be used as a functional equivalent for MCKO
MDO[M:0]	Message Data Out (MDO[M:0]) are output pin(s) used for sending messages such as trace export and other read operations, memory substitution accesses, etc. Depending upon output bandwidth requirements, one, two, four, eight, or more pins may be implemented
MSEO[1:0]	Message Start/End Out (MSEO [1:0]) are output pins that indicate when a message on the MDO pins has started, when a variable-length packet has ended, and when the message has ended. Only one MSEO pin is required, but two pins provide for more efficient transfers
EVTO	Event Out (EVTO) is an optional output pin to development tools indicating exact timing for a single breakpoint status indication. Upon a breakpoint occurrence of the programmed breakpoint source, EVTO is asserted for a minimum of one clock period of MCKO
MCKI	Message Clocking (MCKI) is a free-running input clock from development tools for timing MDI and MSEI pin functions. MCKI can be independent of the embedded processor's system clock.
MDI[N:0]	Message Data In (MDI[N:0]) are inputs used for downloading configuration data, writing to on-chip registers or memory, etc Depending upon input bandwidth requirements, multiple pins may be implemented
MSEI[1:0]	Message Start/End In (MSEI [1:0]) are inputs that indicate when a message on the MDI pins has started, when a variable-length packet has ended, and when the message has ended. Only one MSEI pin is required, but two pin implementations provide more efficient transfers
EVTI	Event In (EVTI) is an input pin allowing off chip control such as processor halts (breakpoints) or synchronized Program/Data Messages
RSTI	Reset In (RSTI) is a pin for resetting the Nexus port resources

Source: Nexus 5001 Forum.

available, when to terminate retrieving a variable-length packet, and when an output public message is ended.

The width of the output message register will be vendor-defined, where the vendor may optimize the register size depending on the size of the packets transmitted. Figure 11.3 shows the state flow for accessing the public message registers as well as other NRRs.

11.3.2 NEXUS AUX Interfaces

The AUX interfaces are unidirectional (either data in or data out), with each AUX port having its own clock. The data out pins of an AUX interface are typically used for trace, and the data in mode is typically used for configuration or calibration of an IC. AUX data in and out ports may be operated concurrently. Nexus also specifies how a JTAG interface can be used in conjunction with the AUX ports. JTAG

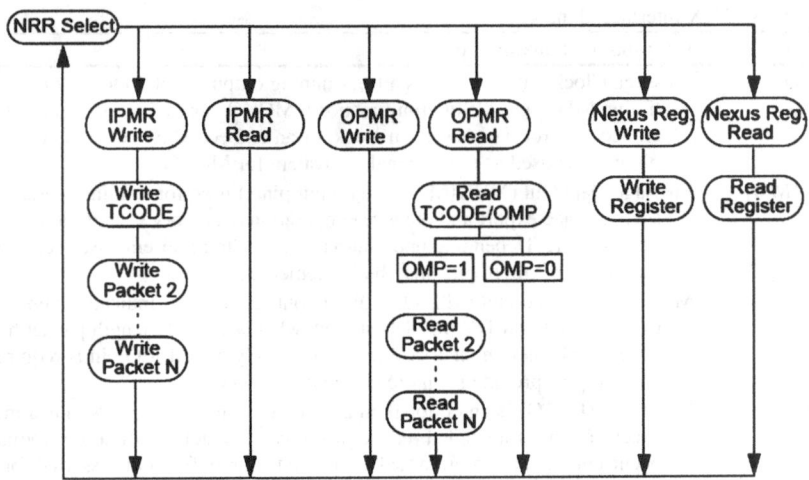

Fig. 11.3 Nexus JTAG message controller state diagram. *Source*: Nexus 5001 forum. All rights reserved

interface operations in Nexus may be used for both configuration and control of the on-silicon instrumentation and for embedding Nexus protocol and data into a JTAG message. Both AUX and JTAG interfaces are controlled by FSM-based controllers, allowing a variety of transfer operations. There are two FSMs for receiving and transmitting messages via the auxiliary pins using the MSEI and MSEO pin functions. A minimum of one and a maximum of two MSEI pins shall provide the protocol for the embedded processor receiving messages, and a minimum of one and a maximum of two MSEO pins shall provide the protocol for the embedded processor transmitting messages.

The Nexus standard defines an extensible auxiliary port that may either be used with the IEEE 1149.1(JTAG) port or as a stand-alone development port. The Nexus standard defines the auxiliary pin functions, transfer protocols, and standard development features to support both 1149.1 and AUX usage. The auxiliary port provides a wider, higher-bandwidth data transfer conduit and can define both AUX input and output ports. Auxiliary out ports are used primarily to provide additional pins in the port for higher throughput for trace output.

For a full-duplex AUX with IEEE 1149.1 pins, a minimum of two auxiliary pins are required for compliance [message data out and message start/end out], assuming a system clock out pin can be used for MCKO. EVTI is also recommended for tool-initiated synchronization. The performance classification, however, would also be minimal and may meet the transfer bandwidth requirements for low-end applications or lower-compliance classifications.

Nexus implementations may have one or two messaging start/end out pins, depending on complexity of the input and output state machines. A two-bit messaging pin allows back-to-back data transfers, speeding delivery of memory data or trace information.

The MSEI/MSEO protocol comprises the following:

- Two "1"s followed by one "0" indicates start of message.
- "0" followed by two or more "1"s indicates end of message.
- "0" followed by "1" followed by a "0" indicates end of variable-length packet.
- "0"s at all other clocks during transmission of a message.
- "1"s at all clocks during no message transmission (idle).

The same sequence is followed when using one or two MSEI/MSEO pins, but when using two MSEI/MSEO pins, it is possible for two sequences to occur on the same clock. MSEI/MSEO is used to signal the end of variable-length packets and not device-specific or fixed-length packets. MSEI/MSEO are sampled on the rising edge of MCKI/MCKO.

Figure 11.4 shows the finite-state machine diagram for one-pin MSEI/MSEO transfers. When using only one MSEI/MSEO pin, the end-message state does not contain valid data on the MDI/MDO pins. Also, it is not possible to have two consecutive end-packet messages. This implies that the minimum packet size for a variable-length packet is two times the number of MDI/MDO pins. This ensures that a false end-of-message state is not entered by transmitting two consecutive 1s on the MSEI/MSEO pin before the actual end of the message.

Systems with class 2, 3, and 4 features primarily use the AUX interfaces, Rules of embedding a Nexus packet in an AUX port are consistent with many other parallel port protocols, with key rules as follows:

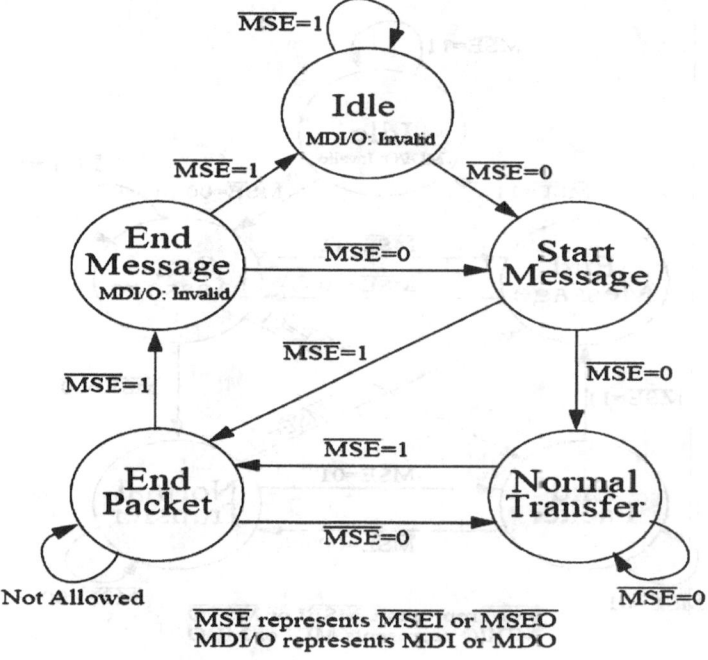

Fig. 11.4 Nexus AUX FSM (single-bit MSE). *Source*: Nexus 5001 Forum.

- A variable-sized packet within a message must end on a port boundary.
- A variable-sized packet may start within a port boundary only when following a fixed-length packet. (If two variable-sized packets end and start on the same clock, it is impossible to know which bit is from the last packet and which bit is from the next packet.)
- Whenever a variable-length packet is sized such that it does not end on a port boundary, it is necessary to extend and zero-fill the remaining bits after the highest-order bit so that it can end on a port boundary. For example, if the MDO port is four bits wide and the unique portion of an indirect address TCODE is five bits, then the remaining three bits of MDO must be packed with 0s.
- A data packet within a data message must be 8, 16, 32, or 64 bits in length.
- To improve message compression, multiple device-specific or fixed-length packets may start and end on a single clock.
- Each type of device-specific or fixed-length packet must be the same within all messages. For example, if a vendor implements three bits to identify the source processor, then all public messages with a source processor packet must be three bits in length.
- When a device-specific or fixed-length packet follows a variable-sized packet, the device-specific or fixed-length packet must start on the port boundary.
- The MSEI/MSEO protocol must be followed for both input and output messages.

Figure 11.5 shows the FSM for two-pin MSEO transfers. The two-pin MSEI/ MSEO option is more robust than the one-pin option. Termination of the current

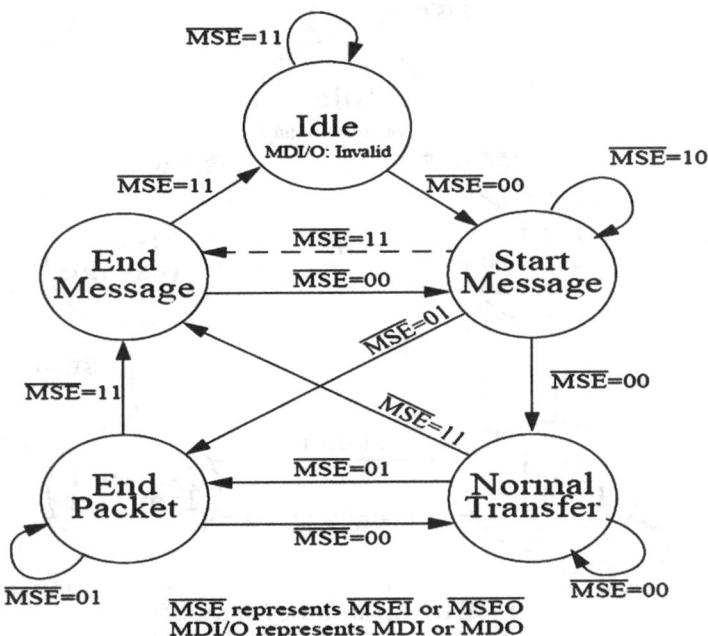

Fig. 11.5 Nexus AUX FSM (2-bit MSE). *Source*: Nexus 5001 Forum. All rights reserved

Table 11.6 AUX Interface for a indirect branch message with 1 bit MSEO

	MDO[3:0]				MSE0[0]	
Clock	3	2	1	0	0	Idle
0	X	X	X	X	1	Idle (or end of last message)
1	T3	T2	T1	T0	0	Start message
2	S1	S0	T5	T4	0	Normal transfer
3	I3	I2	I1	I0	0	Normal transfer
4	I7	I6	I5	I4	1	End packet
5	A3	A2	A1	A0	0	Normal transfer
6	A7	A6	A5	A4	1	End packet
7	X	X	X	X	1	End message
8	T3	T2	T1	T0	0	Start message

Source: Nexus 5001 Forum.

Table 11.7 AUX Interface for a indirect branch message with 2 – bit MSEO

	MDO[3:0]				MSEO[1:D]		
Clock	3	2	1	0	1	0	
0	X	X	X	X	1	1	Idle (or end of last message)
1	T3	T2	T1	T0	0	0	Start message
2	S1	S0	T5	T4	0	0	Normal transfer
3	I3	I2	I1	I0	0	0	Normal transfer
4	I7	I6	I5	I4	0	1	End packet
5	A3	A2	A1	A0	0	0	Normal transfer
6	A7	A6	A5	A4	1	1	End packet/message
7	T3	T2	T1	T0	0	0	Start message

Source: Nexus 5001 Forum.

message may immediately be followed by the start of the next message on the consecutive clocks. An extra clock to end the message is not necessary as with the one-pin MSEI/MSEO option. The two-pin option also allows for consecutive end-packet states. This can be an advantage when small, variable-sized packets are transferred. Tables 11.6 and 11.7 show an examples of data transfer at the AUX interface for the respective cases of one and two bit MSE FSMs.

11.4 Multicore Nexus Debug Approaches

Nexus implementations can support the concurrent debug of both processor and bus operations. Although each processor or logic/bus element in a design may have a native debug environment, debug information can be reformatted using Nexus interface wrappers, which embed debug information into packet fields of the Nexus messages. These Nexus messages can then be merged at a Nexus port control level

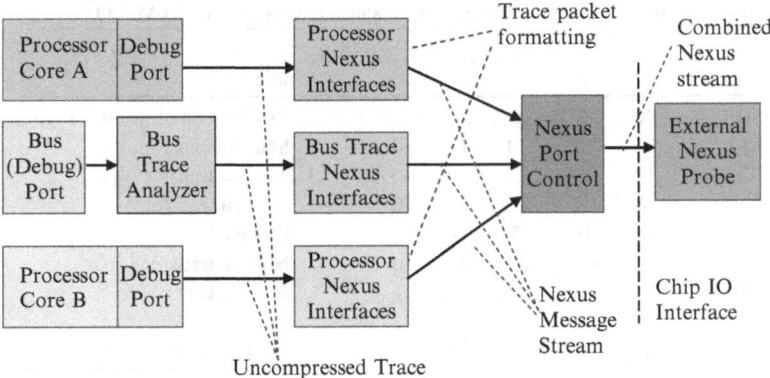

Fig. 11.6 Basic Nexus multicore debug flow

to allow packets from many debug sources to be communicated over a common Nexus port. Because each debug block can be assigned an independent identification (DID) value, debug information can be redirected once off chip, at the probe interface or as a software operation.

Figure 11.6 shows this debug data flow, supporting a multicore architecture consisting of two processor (or other) cores and a bus port or other bus-level debug interface. All blocks have some native debug or analyzer blocks. The debug information is made into Nexus-compliant messages, including any additional compression, by in-line Nexus interface blocks with the different independent message streams consolidated into a single combined Nexus stream at the port interface.

One of the issues in debug of multiple core systems is that even when debug information from different blocks is combined into a single Nexus stream, the control and synchronization of debug over many different core or subsystems remains largely independent. Having better control and synchronization of different debug resources can significantly improve debug efficiency. In addition to the Nexus interfaces for each of the processor on-chip debug resources, the environment includes Nexus-controlled cross-triggering and systemwide timestamping resources to help synchronize and cross-reference debug operations occurring at different parts of the architecture, allowing different off-chip debugger environments to better comprehend the context and operations occurring in other parts of a design.

Nexus provides a toolbox and an approach to implementing a debug architecture, which can be customized to properly address different architectures and unique analysis considerations. Properly implemented, a comprehensive debug solution can measurably improve the level of testability, maintainability, and analysis capabilities throughout the life cycle of a chip design, but implementing the correct on-chip debug solutions also requires an engineering investment in understanding how debug tools will be used as well as the considerations of all the trade-offs for integrating debug solutions into a design Fig. 11.7.

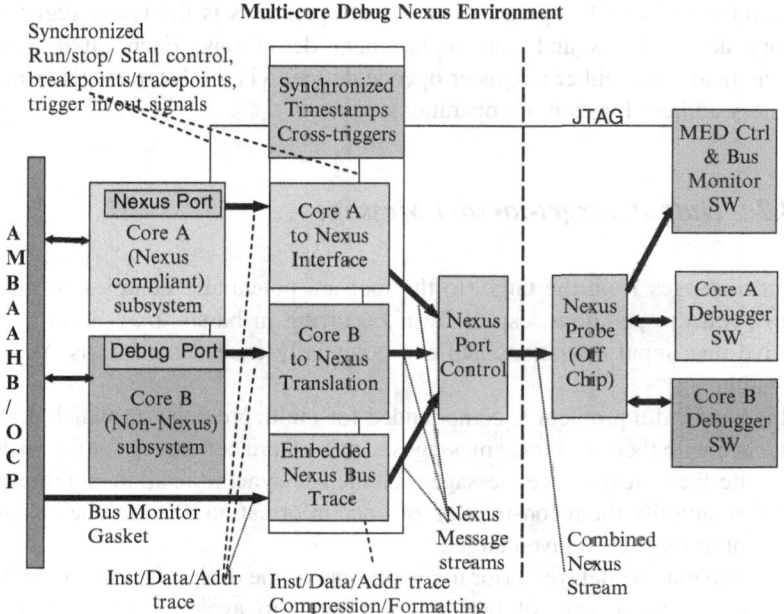

Fig. 11.7 Multicore debug Nexus environnent

Nexus allows embedded processor implementations that comprise multiple clients to use a single AUX, depending on the transfer bandwidth requirement for the application. The AUX may be designated for a single client or shared by multiple clients on the embedded device during runtime. Messages transmitted via the AUX contain information defined by the Nexus standard indicating which client generated the message. Implementations can include clients on a single chip (Fig. 11.8) as well as processors on multiple chips sharing a Nexus interface.

Because the transfer of information is message-based, a variety of scheduling and transfer methods of simply parsed and disassembled messages between the Nexus interfaces and different cores are supported, allowing delayed and prioritized transfer of information between several cores and the Nexus interfaces. Because their characteristics differ, we shall consider the cases of AUX in (from tool to target) and AUX out (from target to tool) messages separately.

11.4.1 Input Tool-to-Target Messages

Managing Nexus input messages in a multicore system is straightforward, because there is typically a single host generating messages over the debug interface and only one message will be queued for transfer to the on-chip target at any given time.

The number of TCODE operations for input operations is limited to register and memory access types and port replacement definitions. Each input message contains fields with either a register opcode defined via the Nexus register map or a memory address for memory operations.

11.4.2 Output Target-to-Tool Messages

Output messages from the target to the tool are potentially complex to manage, because (trace) operations, especially if occurring in bursts, may be more data-intensive than input operations and can potentially exceed the Nexus AUX port bandwidth.

This bandwidth problem is compounded for multicore debu, in which different cores, each with their own trace messages to export, are competing for access to the Nexus interface. Nexus trace messages can include synchronization and timestamp fields that simplify the reconstruction of trace information that may be delayed in being sent to tools for a given target.

If a trace may be delayed prior to export, one of the design factors in the Nexus blocks should be a level of buffering sufficient to avoid dropping or loss of messages while waiting for access to the AUX out port. There are a variety of ways to manage output data from multiple sources. A simple approach is to configure a simple static output multiplexor to choose between different Nexus message streams and disable Nexus traffic for the duration of the sub-system not chosen. If this duration is significant or is competing with other data-intensive messages (memory access, for example), this can result in the need for larger on-chip buffers to avoid losing trace messages.

The nature of multicore systems analysis, however, is that for many problems, debug requires access to concurrent information from several cores in order to sufficiently understand the issues involved. To avoid the need for large on-chip buffers, more sophisticated message control can be implemented to provide scheduling, prioritization, and arbitration of Nexus messages.

Nexus messages can be merged at a Nexus port control level to allow packets from many debug sources to share a common Nexus port. Because each debug block can be assigned an independent identification value, debug information can be redirected once off-chip at the probe interface or as a software operation.

The packet nature of Nexus messages allows a variety of network queuing techniques to interleave messages from multiple sources into a common AUX out port. The intelligence for this may be implemented in on-chip controller hardware with different implementations based on output multiplexing, enable logic, and a funneling logic shown in Figs. 11.8–11.10, or in off-chip software with priorities transferred to a simpler AUX out control block as Nexus input messages.

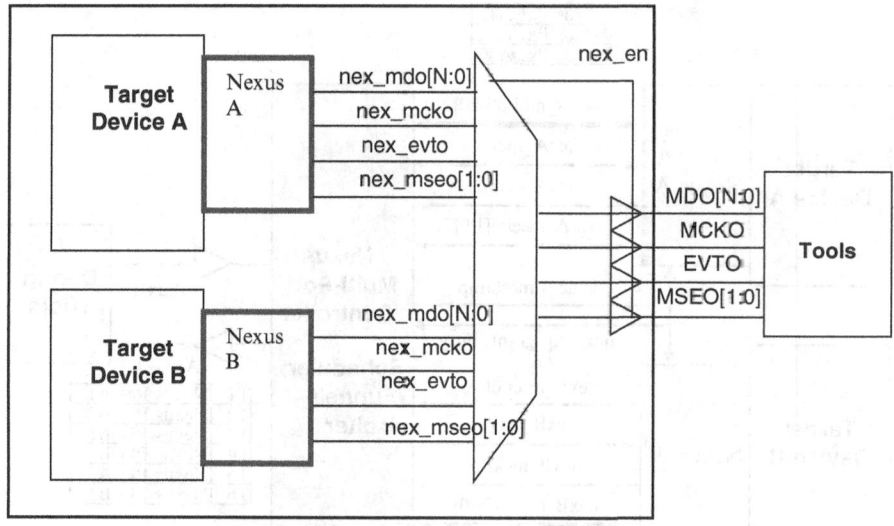

Fig. 11.8 Multicore single-chip AUX multiplexor interface

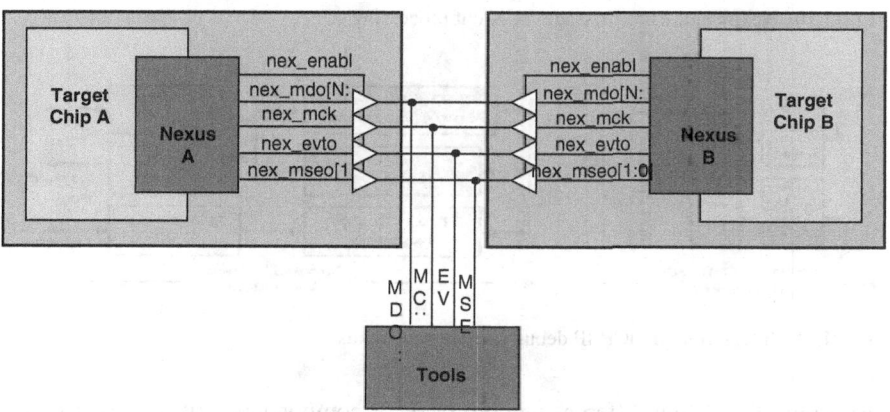

Fig. 11.9 Multichip AUX interface

11.5 Nexus Product Implementations

In 2007, the Nexus Forum and OCP-IP developed a collaborative agreement on debug sub-systems. A reference design for attaching a Nexus port and communicating with OCP-IP debug signals was developed (Fig. 11.11) and is included as an appendix to the OCP-IP debug working group specification. The key element of the interface is use of the ownership trace message and source fields defined for each

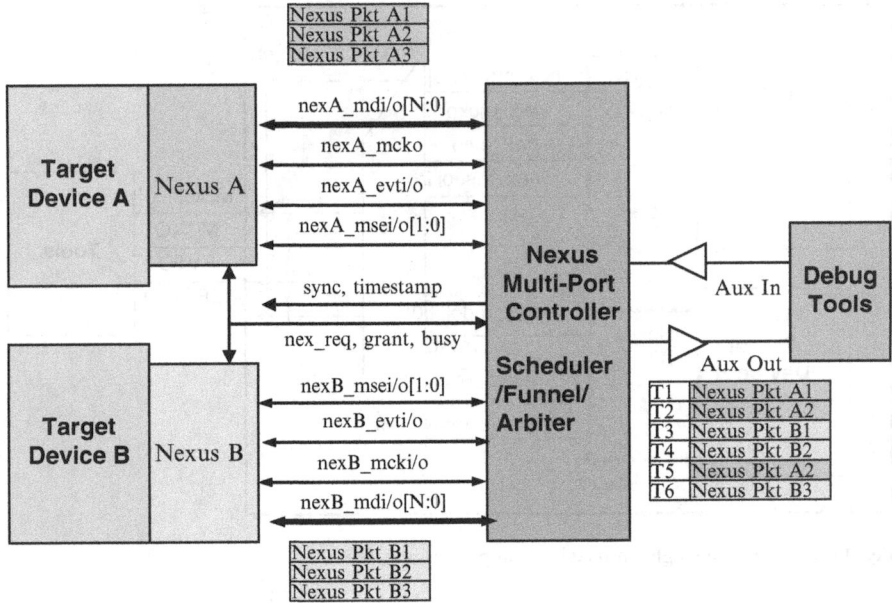

Fig. 11.10 Nexus multicore message AUX out processing

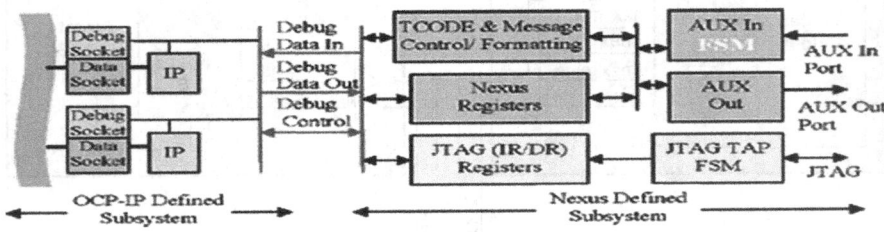

Fig. 11.11 Integration of OCP-IP debug sockets and Nexus

instruction that allow different on-chip OCP-IP components with different debug sockets to be accessed concurrently.

Semiconductors from Freescale have been built and implemented for the majority of Nexus-based SoCs. These SoCs have serviced many industrywide markets, including automotive, wireless, and networking. Two example of Freescale SoCs using Nexus are discussed in this section.

One family of SoCs, initially offered for the automotive powertrain market, uses the multiprocessing features of Nexus to provide debug visibility to the processor core – a PowerPC e200z6, the enhanced timer processor units (ETPU), and the secondary peripheral bus.

The MPC5500 family of SoCs support various debug facilities as shown in Fig. 11.12. There are five major architectural blocks that provide the debug functionality:

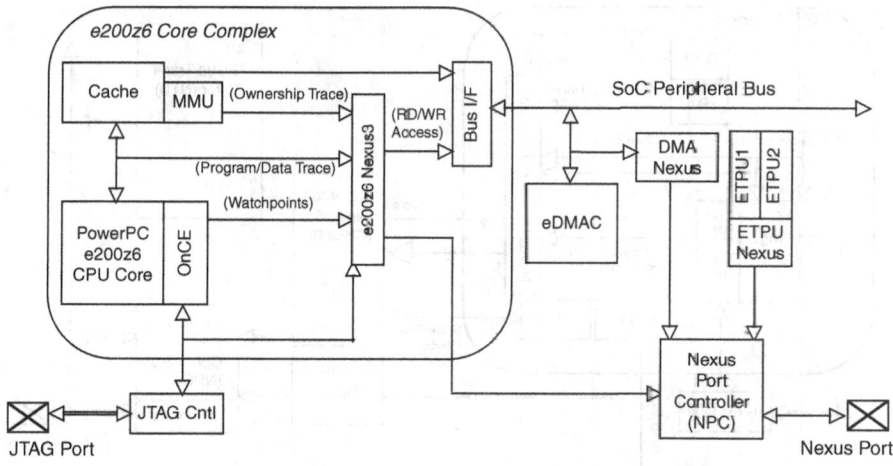

Fig. 11.12 Freescale MPC5500 multi-core Nexus implementation

- PowerPC e200z6 Nexus1 module (OnCE) – Class 1–compliant debug of the processor.
- PowerPC e200z6 Nexus3 module – Class 3–compliant trace of the processor.
- DMA Nexus module – Data trace support for DMA data access.
- ETPU Nexus – Class 3–compliant trace of enhanced timer processor units.
- Nexus port controller – Arbitration for Nexus I/O port.

The PowerPC e200z6 Nexus module supports Nexus Class 1 and Class 3 features as well as the optional features of watchpoint trigger enable of program/data tracing and burst capability on Nexus-initiated read/write accesses.

Class 1 features such as breakpoint generation, single stepping, and internal resource access (processor halted) are handled within the processor's JTAG-based static debug OnCE (a Freescale proprietary on-chip emulation) block. Watchpoints for Nexus3 are also generated within the OnCE module. These eight watchpoints (for various programming events) can be used to trigger trace-enable/disable, generate watchpoint messages, and drive an optional EVTO output pin.

The DMA Nexus module supports tracing data reads and writes on the peripheral bus. The Nexus port controller (NPC) module arbitrates between the various debug modules for the shared port and controls the port settings (MCKO divide ratio, port-width option).

The second example is from a family of wireless baseband processors nicknamed mxC (Fig. 11.13). The first generation of these SoCs combines a StarCore SC1400 DSP with an ARM11xx core and various mixes of peripherals and memory configurations.

The DSP sub-system supports a slightly more enhanced set of debug facilities. The major architectural blocks consist of:

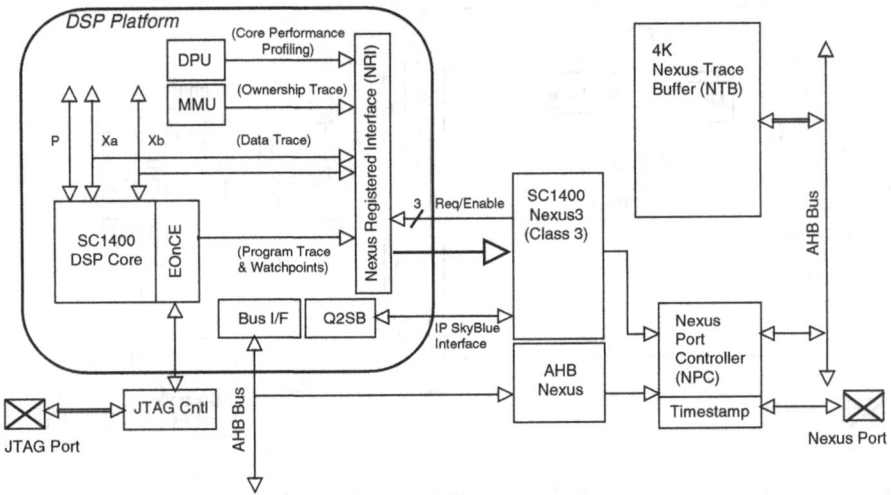

Fig. 11.13 Freescale mxC DSP sub-system and multi-core Nexus implementation

- SC1400 Nexus1 module (EOnCE) – Class 1–compliant debug of the DSP.
- SC1400 Nexus3 module – Class 3–compliant trace of the DSP.
- AHB Nexus module – Data trace support for AHB data access.
- Nexus trace buffer – Shared internal memory for dumping Nexus trace data.
- Nexus port controller – Arbitration for Nexus I/O port and timestamp generator.

The SC1400 Nexus modules also support Nexus Class 1 and Class 3 features as well as the optional features of watchpoint trigger-enable of program/data tracing and data-acquisition messaging for data logging. In addition, the Nexus3 module supports vendor-defined triggering of program/data tracing using the process ID, and specific messages for reporting core performance profiling information from the SC1400 debug and profiling unit (DPU).

Class 1 features such as breakpoint generation, single stepping, and internal resource access (processor halted) are handled within the processor's JTAG-based static debug block, EOnCE. Watchpoints for Nexus3 are also generated within the EOnCE module. These seven watchpoints (for various programming events) can be used to trigger trace enable/disable and generate watchpoint messages, and can be connected to a cross-triggering module for triggering events in other portions of the SoC. They also drive an optional EVTO output pin.

The AHB Nexus module supports tracing data reads and writes on the peripheral bus and can generate additional watchpoints based on AHB address and/or data values. These watchpoints can also be used by a cross-triggering module within the SoC. Additional AHB Nexus modules support data trace on the application side (ARM11) of the baseband as well.

Similar to the MPC5500 family, the Nexus port controller module arbitrates between the various debug modules for the shared port. In addition to the arbitration and port control, the mxC NPC module provides timestamping capability for the debug system by maintaining an "absolute" timestamp value that the individual Nexus modules can use within their messages or to generate their own "relative" timestamp to reduce bandwidth penalty.

The mxC SoCs also support internal storage of Nexus messages to an internal Nexus trace buffer (NTB) for retrieval at a later time. These messages are sent to AHB memory within the SoC, which allocates a secondary function for the storage of trace information. This information can be read out through the JTAG port (or other memory-access mechanism) when real-time visibility is not as critical. This allows more trace data to be stored by reducing bandwidth restrictions associated with sending data off-chip.

11.6 Summary

Nexus has been evolving as an IEEE standard for several years and is seeing increased use as a debug solution in many different architectures and markets. Using Nexus provides several advantages to designers, providing a widely supported infrastructure and a framework for customized solutions. As an "architecture-agnostic" interface, Nexus also provides advantages to tool vendors by reducing development costs and time to market. Freescale has been an industry leader in developing Nexus-based solutions to support a range of processor cores and configurations. The technical committee within the IEEE-ISTO 5001 consortium is continually working to add feature enhancements to the standard and support for a wider range of SoC architectures.

Chapter 12
MIPS EJTAG

EJTAG is a hardware/software sub-system that provides comprehensive debugging and performance-tuning capabilities to MIPS® Technologies–developed processors and to SoC components with MIPS processor cores. Like many other processor debug blocks, it uses the IEEE 1149.1 JTAG TAP as an external interface. Higher-performance debug can also use a complementary parallel port (PDtrace) for data transfers.

EJTAG is tightly coupled to the MIPS instruction set and is typically packaged as part of a MIPS processor license as an optional resource architecture for processor and system debugging Fig. 12.1. The MIPS architecture has historically provided a set of primitives for debugging software, which includes:

- A breakpoint instruction, BREAK, whose execution causes a specific exception.
- A set of trap instructions, whose execution causes a specific exception when certain register value criteria are satisfied.
- Dual optional watch registers that can be programmed to cause a specific exception on a load, store, or instruction fetch access to a specific 64-bit double word in virtual memory.
- An optional TLB-based MMU that can be programmed to trap on any access or, more specifically, on any store to a page of memory.

ETJAG has evolved, and there is limited backward compatibility between the current 5.x version and earlier versions, especially for revisions 2.5 and earlier. This is sometimes a problem because different processors may use different versions of EJTAG. As a prominent example, the Toshiba TX series of MIPS processors use a circa 2.5 level of EJTAG that diverges sufficiently from current EJTAG in terms of register usage and naming, as well as changes in debug instructions and other features, and so must be considered semi-independently.

The EJTAG registers are generally 32 bits wide for MIP32 architectures and 64 bits wide for MIPS64 architecture, so specific bit mappings depend on implementation. Registers set up the debug resources and capture debug status information during the debug operation. Registers are memory-mapped and accessible from the EJTAG probe. Operation of the EJTAG circuitry is controlled through an EJTAG probe that interfaces the host development system and the target device. There are numerous probes available that support EJTAG.

N. Stollon, *On-Chip Instrumentation: Design and Debug for Systems on Chip*,
DOI 10.1007/978-1-4419-7563-8_12, © Springer Science+Business Media, LLC 2011

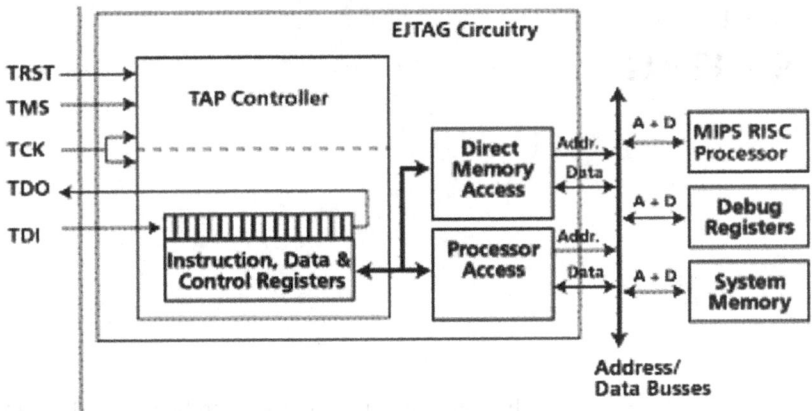

Fig. 12.1 General overview of the EJTAG interface. *Source*: MIPS Technologies, Inc. All rights reserved

EJTAG processor core extensions are required in any EJTAG implementation; many of the features are implementation-optional. Information on which EJTAG features are implemented is found in the DCR register:

- The single-step execution feature is optional. The presence or absence of single-step execution capability is indicated to debug software via the CP0 debug register.
- The debug interrupt request from the TAP via the DINT probe signal or through an implementation-dependent internal signal is optional.
- The TAP is optional.
- The hardware breakpoint unit (HBU) is optional.
- The debug control register is optional. Note that it is required if either the TAP or the HBU is implemented.

The processor access and DMA circuit blocks are used to set up and monitor the processor's internal buses and to execute the code from the EJTAG interface. In order to provide debug code without integrating it into the application code, the EJTAG processor-access circuitry shares a specific memory location that can replace system memory in debug mode. When the processor accesses this memory space, the EJTAG circuitry can feed it debug instructions not present in the application code.

When an access is detected, the EJTAG circuitry makes the transaction address available in the EJTAG address register. The appropriate data is also made available in the EJTAG data register if the operation is a write, and it is inserted into the EJTAG data register if the operation is a read.

The EJTAG DMA circuitry enables the EJTAG to initiate transactions on the system bus while running application code, providing access to debug and user memory areas. This makes it possible to inspect debug resources and user memory while the system is executing its code, providing excellent visibility into system operation with little or no impact on real-time operation. Setup of DMA activities is handled by setting up the EJTAG registers. Using the DMA access circuitry, it is possible to download application code or transfer user memory off-chip while the debug session is ongoing.

The EJTAG debug features require high integration with the processor. Different generations of MIPS processors have differences in debug modes, registers, and instructions to support the debug process.

12.1 EJTAG Instructions and Registers

EJTAG provides a standard debug I/O interface, enabling the use of traditional MIPS debug facilities on SoC components. In addition, EJTAG provides the TAP instructions that allow access to corresponding EJTAG registers, for the following:

- IDCODE: Device identification register with manufacturer, part number, and version ID for the specific chip (IR 00000001).
- IMPCODE: Implementation register indicating implemented EJTAG features in this specific chip (IR 00000011).

ADDRESS EJTAG: Address register used to access the on-chip address bus (IR 00001000).

- DATA EJTAG: Data register used to access the on-chip data bus (IR 00001001).
- CONTROL EJTAG: Control register used for setup and status information (IR 00001010).
- ALL: Access to EJTAG address, data, and control registers in one chain (IR 00001011).
- EJTAGBOOT: Causes processor reset followed by a debug exception (IR 00001100).
- NORMALBOOT: Causes processor reset followed by execution of the reset handler (IR 00001101).
- FASTDATA: Provides a one-bit tag in front of the data register to capture the processor access pending bit for fast data transfer; access to the data and FastData registers (IR 00001110).
- TCBCONTROLA: Access to the control register TCBControlA in the trace control block (TCB); used by external probe (debugger) software to control tracing output from the core (IR 00010000).
- TCBCONTROLB: Access to the other control register TCBControlB in the TCB that controls tracing configuration options (IR 00010001).
- TCBDATA: Provides access to the registers specified by the TCBCONTROLB REG field (IR 00010010).
- TCBCONTROLC: Access to the control register TCBControlC in the TCB and used in the TCB; controls tracing configuration options (IR 00010011).
- PCSAMPLE: Access the PC sample register (IR 00010100).
- BYPASS: One-bit register with no operation; JTAG required (IR 1111111).

The size of the EJTAG address and data registers depends on the specific implementation, but they are usually at least 32 bits. The size of the device ID, implementation, and EJTAG control registers is 32 bits; these registers allow the user to perform debug setup and provide important status information during the debug session.

The processor's memory-mapped EJTAG memory is located in the debug memory segment, which is a sub-segment of the debug segment. It is accessible by debug software when the processor is executing in debug mode. An EJTAG probe handles all access to this segment through the TAP; the processor has access to dedicated debug memory even if no debug memory was originally located in the system.

To allow inspection of the processor state at any time in the execution flow, a debug exception with priority over all other exceptions is introduced. When a debug exception occurs, the processor goes into debug mode, where it has unrestricted access to coprocessors, memory areas, etc.

The debug exception handler is executed in debug mode and provided by the debug system. It can be executed from the probe through a processor access, or it may reside in the application code if the developer chooses to include a debug task in the application. An overall requirement is that debugging be nonintrusive to the application so that execution of the application can be resumed after the required debug operations are run. However, loss of real-time operation is inevitable when the debug exception handler is executed. Even if other parts of the system are halted during the debug operations.

The debug control register controls and provides information about debug issues. The width of the register is 32 bits for 32-bit processors and 64 bits for 64-bit processors. The DCR provides the following key features:

- Interrupt and NMI control when in nondebug mode.
- NMI pending indication.
- Availability indicator of instruction and data hardware breakpoints.
- Availability of the optional PC sample feature and the sample period being used.

For EJTAG features, there is no difference between a reset and a soft reset occurring to the processor; they behave identically in both debug mode and nondebug mode.

Data hardware breakpoint registers are controlled as memory-mapped registers. Most registers have separate instances for each implemented data hardware breakpoint, as indicated with an "n" in the following list.

Data Hardware Breakpoint Registers

Register name	Register	Mnemonic functional description
Data Breakpoint Status	DBS	Indicates number of data hardware breakpoints and status on a previous match
Data Breakpoint Address(n)	DBAn	An address to compare for breakpoint n
Data Breakpoint Addr Mask(n)	DBMn	Address comparison mask for breakpoint n
Data Breakpoint ASID(n)	DBASIDn	ASID value to compare for breakpoint n
Data Breakpoint Control (n)	DBCn	Control of breakpoint n: match on load/store, data bytes, access to data bytes, comparison of ASID, and generated event on match
Data Breakpoint Value (n)	DBVn	Data value to match for breakpoint n

12.2 PC Sampling

PC sampling is one of the unique optional features of EJTAG, used for program profiling and analysis; it samples the value of the PC periodically. This information can be used for statistical profiling of the program, akin to gprof. This information is also very useful for detecting hot spots in the code. In a multithreaded environment, this information can be used to detect thread behavior and verify thread scheduling mechanisms in the absence of the PDtrace facility.

The presence or absence of the PC sampling feature is available in the debug control register. If PC sampling is implemented, then the PC values are constantly sampled at the requested rate. The sampled PC values are written into a TAP register. The old value in the TAP register is overwritten by a new value even if this register has not been read out by the debug probe. The sample rate is specified by a field in the debug control register called PCSR (PC sample rate). Note that the processor samples PC even when it is asleep, that is, in a WAIT state. This permits analysis of the amount of time spent by a processor in the WAIT state, which is important to understand in real-time and power sensitive applications.

The sampled values include a new data bit, the PC, the ASID (address space identifier, a MIPS-tagged TLB) of the sampled PC, as well as the thread context ID if the processor implements the MIPS MT ASE. The new data bit is used by the probe to determine if the PCsample register data just read out is new or has already been read and can be discarded.

The sampled PC value is the PC of the completing instruction in the current cycle. If the processor is stalled when the PC sample counter overflows, then the sampled PC is the PC of the next completing instruction. The processor continues to sample the PC value even when it is in debug mode.

12.3 MIPS PDtrace™

The tracing logic within the processor core outputs all trace information on the PDtrace™ interface. This PDtrace interface connects to the on-chip TCB unit. The TCB is responsible for collecting the trace data sent every cycle on the PDtrace interface by the core's tracing logic. The TCB captures and stores this trace data, in different configurations, in either an on-chip trace memory or an off-chip trace memory using the probe.

Figures 12.2–12.4 show some of the different configurations for the PDtrace. A Probe Interface block (PIB), which communicates between the trace-related blocks, and external interfaces is used only for dedicated trace interfaces and is not needed for trace exported through the EJTAG port. The key blocks in the PD trace subsystems are

- The TCB, which provides temporary on-chip storage of trace information.
- The interface between the TCB and the TAP controller.
- The PIB.
- The external probe interface.

The TCB can be configured for three primary interfaces:

- The PDtrace interface to the processor core.
- The TCB TAP interface, which connects the EJTAG TAP controller resident within the processor core to the TAP functionality present within the TCB.
- An optional TCtrace interface to the PIB.

One main function of the TCB is to capture trace information from the PDtrace interface and store it to trace memory. This trace information is then analyzed by the trace reconstruction software in the debugger. Because tracing the entire run of a program can require large amounts of storage, compression of trace information is desirable. Although the trace information undergoes one level of compression in the core, further compression is possible before the trace information is stored to trace memory by the TCB. The TCB achieves this compression using a number of trace formats that eliminate the storage of unnecessary trace bits in each cycle. This section describes these formats.

Figure 12.2 shows the TCB, the PIB, and the trace data path from the PDtrace IF to the Probe IF. It is optional whether the TCB implements on-chip trace memory and/or the TCtrace IF with a PIB and off-chip trace memory. Figure 12.3 shows the TCB streaming data to off-chip trace memory through the PIB. The number of pins needed for trace data on the probe IF is configurable to 4, 8, or 16.

Figure 12.4 shows a configuration in which the TCB is streaming data to on-chip trace memory. The size of the on-chip trace memory is configurable. After trace capture has stopped, the trace data in the on-chip memory is accessed through the EJTAG probe.

12.3.1 Trace Output Formats

The amounts of Trace information that is exported varies significantly based on the types of instructions and operations being processed. As discussed previously, unless there are conditional operations or discontinuities in the processor operation, instruction trace can be reconstructed based on minimal information base on the assumptions of normal operational flow. If there is an instruction discontinuity, as example a branch, a jump, or an interrupt, the amount of trace information required increases. In a MIPS processor trace blocks, this is performed by configuring the trace into one of six different trace packet formats based on the amount of information that is required. Trace may be performed on either single-pipeline or multiple pipelines. For simplicity, we will only discuss the formats when the core being traced is in a single-pipeline or single-issue implementation. A processor with multiple pipelines requires data synchronization and combining for sending trace information to trace memory. The TCB can perform this combining and formatting to reduce the number of bits that are sent out each cycle. If there are K pipelines within the core, 1, 2, ...K, then for each cycle, the TCB generates a trace format from each pipeline, in the respective order.

Trace format 1 (TF1): When the processor is stalled, no execution trace information needs to be recorded except that this was a stall cycle. This is done using a

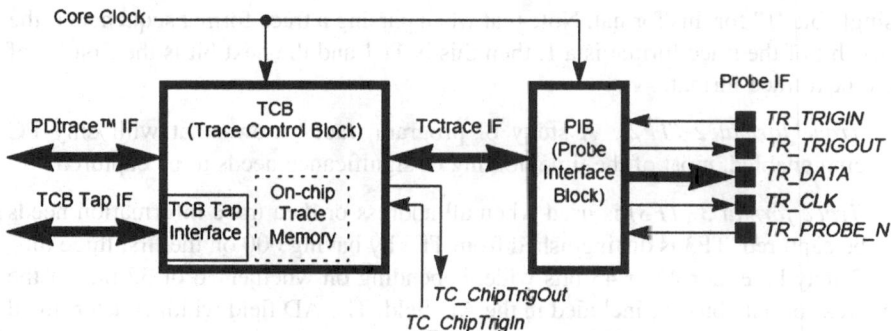

Fig. 12.2 PDtrace interfaces. *Source*: MIPS Technologies, Inc. All rights reserved

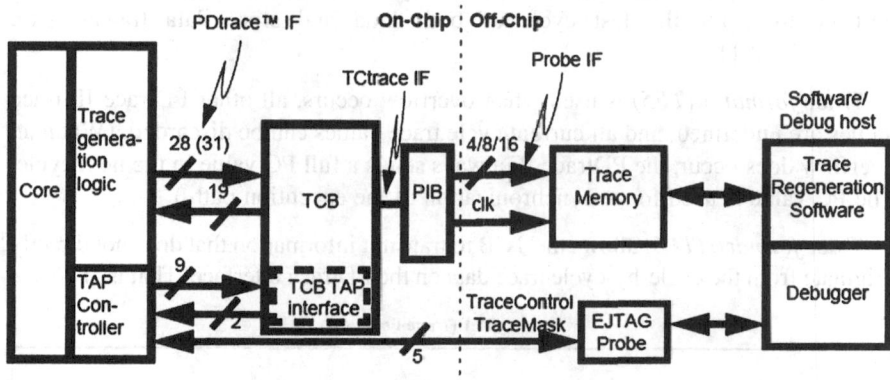

Fig. 12.3 PDtrace off-chip streaming interface. *Source*: MIPS Technologies, Inc. All rights reserved

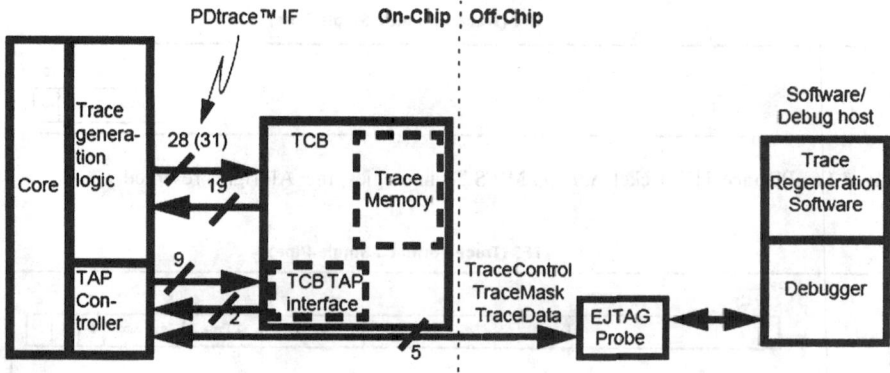

Fig. 12.4 PDtrace on-chip streaming interface. *Source*: MIPS Technologies, Inc. All rights reserved

single bit "1" for this format. Note that when parsing a trace format sequence, if the first bit of the trace format is a 1, then this is TF1 and the next bit is the first bit of the next trace format.

Trace format 2 (TF2): A study of program traces shows that with only PC tracing enabled, most of the time nothing of significance needs to be captured.

Trace format 3 (TF3) is used when all address or data trace information needs to be captured. TF3 is distinguished from TF2 by having 000 on the first three bits. TF3 may be either 27 or 43 bits wide, depending on whether 16 or 32 bits of the address or data bus are included in the AD field. The AD field width is determined by fields in the TCBCONTROLA register.

Trace format 4 (TF4) is the last cycle of a current data transmission. When capturing the cycle-by-cycle values on the PDtrace IF, the last cycle of a load data transmission cannot be distinguished from the last cycle of a store data transmission (without saving information from a previous cycle). This means that the TF4 format will be used for the last cycle of both load and store data transmission Figs. 12.5–12.11.

Trace format 5 (TF5) is used when overflow occurs, all other PDtrace IF trace values are undefined, and all current cycle trace values can be discarded. (When an overflow does occur, the PDtrace IF always sends a full PC value in the next cycle. The PC value is used for re-synchronization of the execution path.)

Trace format 6 (TF6) allows the TCB to transmit information that does not directly originate from the cycle-by-cycle trace data on the PDtrace interface. That is, TF6 can

TF1 (Trace Format 1)

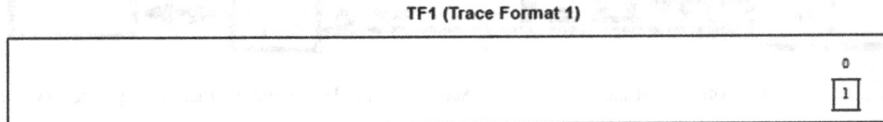

Fig. 12.5 PDtrace TF1 packet. *Source*: MIPS Technologies, Inc. All rights reserved

TF2 (Trace Format 2 Single-Pipe)

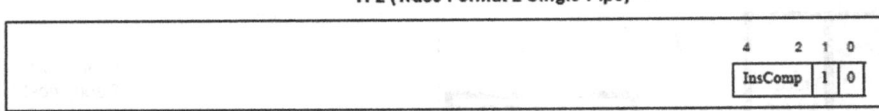

Fig. 12.6 PDtrace TF2 packet. *Source*: MIPS Technologies, Inc. All rights reserved

TF3 (Trace Format 3 Single-Pipe)

26(42)					11	10	9	8		6	5		3	2	1	0
		AD				TMode	TEnd	TType			InsComp		0	0	0	

Fig. 12.7 PDtrace TF3 packet. *Source*: MIPS Technologies, Inc. All rights reserved

TF4 (Trace Format 4 Single-Pipe)

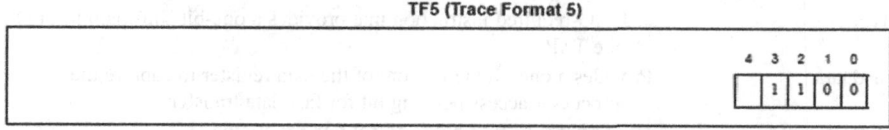

Fig. 12.8 PDtrace TF4 packet. *Source*: MIPS Technologies, Inc. All rights reserved

TF5 (Trace Format 5)

			4	3	2	1	0
			1	1	0	0	0

Fig. 12.9 PDtrace TF5 packet. *Source*: MIPS Technologies, Inc. All rights reserved

TF6 (Trace Format 6)

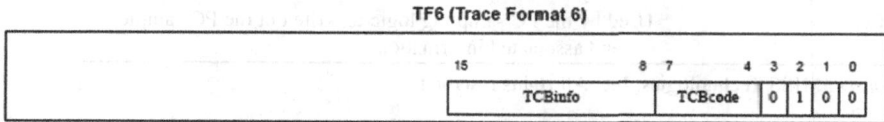

Fig. 12.10 PDtrace TF6 packet. *Source*: MIPS Technologies, Inc. All rights reserved

								Trace											Type	
	5	5	5	4	4	4	3	3	2	2	2	1	1							
TW	9	6	2	8	4	0	6	2	8	4	0	6	2	8	4	0	3	0
1	1	TF2			TF3						TF3									1
2	TF2	1	s	s	s	s	s	s	s	s	TF3		1	TF2	TF2	TF2	1	TF2	TF2	1	1	1	1
3		TF3				1	s	s	s	s	TF3			TF2		TF2		TF2	u	TF2	2
4	u	u	u	u	u	u	u	u	u	u	u	u	u	u	u	u	u	u	u	u	u	u	u	u	u	u	u	u	u	u	u	u	TF6 (stop)			TF3	2

Fig. 12.11 Trace word with TF1 from the sequence in Table 12.2. *Source*: MIPS Technologies, Inc. All rights reserved

be used by the TCB to store any information it wants in the trace memory, within the constraints of the specified format. This information can then be used by software for any purpose. For example, TF6 can be used to indicate a special condition, trigger, semaphore, breakpoint, or break in tracing that is encountered by the TCB.

After compression of data into the trace formats, the trace information must be streamed to either on-chip or off-chip dedicated trace memory. Because each of the major trace formats is a different size, this complicates the efficient storage of this information in fixed-width on-chip memory and the transmission of this data through a fixed-width interface to off-chip memory. To simplify the memory overhead and pin bandwidth issues, the trace formats are first gathered into trace words of regular width Table 12.1.

A TraceWord (TW) is defined to be 64 bits wide. It has a 4-bit type indicator on bits [3:0], and regular TFs stacked up in the remaining 60 bits of the word. The trace portion of a TW consists of one or more trace formats, TF1 through TF6. Note that trace formats TF1, TF2, TF5, and TF6 have fixed size, whereas TF3 and TF4 can vary in size.

Table 12.1 EJTAG registers

EJTAG register	Description
Device ID	Identifies device and accessed processor in the device
Implementation	Identifies main debug features implemented and accessible via the TAP
Data	Data register for processor access used to support the EJTAG memory
Address	Address register for processor access used to support EJTAG memory
EJTAG control (ECR)	Control register for most EJTAG features used through the TAP
Bypass	A JTAG-required instruction that provides a one-bit shift path through the TAP
FastData	Provides a one-bit tag in front of the data register to capture the processor access pending bit for fast data transfer
TCBControlA	Used by the TCB to hold control bits for tracing
TCBControlB	Used by the TCB to hold control bits for tracing
TCBData	Used by the TCB to access data from on-chip trace memory if present
TCBControlC	Used by the TCB to hold control bits for tracing
PCsample	Used by the PC sampling logic to write out the PC sample and associated information

Source: MIPS Technologies, Inc. All rights reserved

Cycle-inaccurate trace: The TF1 format is needed only when a sequence of the trace must show cycle-by-cycle behavior of the processor without missing any cycles. When the trace regeneration software only needs to show the sequence of instructions executed, the TF1 format that shows processor stall cycles can be omitted.

In this situation, additional optimization removes bit zero on the other TFs before storing to trace memory. The example trace sequence in Table 12.2 will then produce the TWs shown in Fig. 12.12. Note that to reconstruct the trace accurately, external software must know what type of tracing was enabled at the TCB.

On-chip trace memory format: The on-chip trace memory is defined to be a 64-bit-wide memory. The TWs are stored in consecutive address locations. The trace memory is only written when a full TW is available, hence a new TW might not be written each cycle because a new TW might not be created each cycle Fig. 12.13.

12.3.2   Trace Control Block Registers

TCBCONTROLA is a control register in the TCB that is mainly used to control the trace input signals to the core on the PDtrace interface. Trace output from the processor on the PDtrace interface can be controlled by the trace input signals to the processor from the TCB. The TCB uses a control register, TCBCONTROLA, whose values are used to change the signal values on the PDtrace input interface. External software (i.e. debugger) can therefore manipulate the trace output by writing to the TCBCONTROLA register. The TCBCONTROLA register is written by an EJTAG TAP controller instruction, TCBCONTROLA.

Table 12.2 An example trace sequence

Cycle #	Trace format	Cycle #	Trace format
1	TF3 (16 significant AD bits)	2	TF3 (16 significant AD bits)
3	TF2	4	TF1
5	TF1	6	TF1
7	TF1	8	TF2
9	TF2	10	TF1
12	TF2	11	TF2
13	TF2	14	TF1
15	TF3 (5 significant AD bits)	16	TF1
17	TF2	18	TF2
19	TF2	20	TF2
21	TF3 (11 significant AD bits)	22	TF1
23	TF3 (6 significant AD bits)	24	TF6 (Stop indicator)

	Trace														Type
	5	5	5	4	4	4	3	3	2	2	2	1	1			
TW	9	6	2	8	4	0	6	2	8	4	0	6	2	8	4 0	3 0
1	TF2	TF2		TF3					TF3							1
2	TF3	TF2	TF2	TF2	TF2	ssssssssssss		TF3			TF2	TF2	TF2	TF2		1
3	TF6 (stop)		ssssssssss	TF3				s	TF3							6
4	uuuuuuuuuuuuuuuuuuuuuuuuuuuuuuuuuuuuuuuuuuuuuuuuuuuuuuuuuuuuuuuuuuuuuuuuuuuuuuuuuuuuTF6															2

Fig. 12.12 Trace word without TF1 from the sequence in Table 12.2. *Source*: MIPS Technologies, Inc. All rights reserved

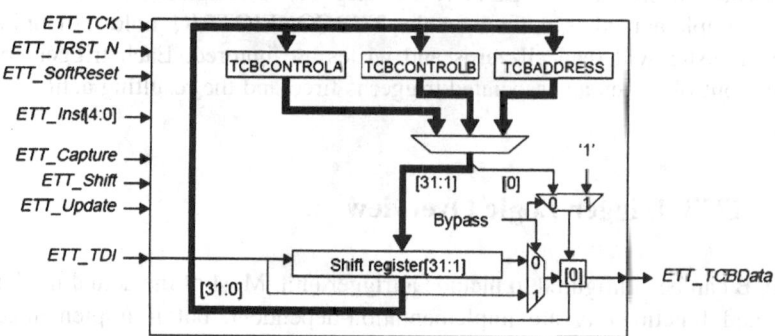

Fig. 12.13 TCB TAP interfaces. *Source*: MIPS Technologies, Inc. All rights reserved

TCBCONTROLB is a control register in the TCB that is mainly used to specify what to do with the trace information. The REG field in this register specifies the number of the TCB internal register accessed by the TCBDATA register. TCBDATA accesses registers specified by the REG field in the TCBCONTROLB register.

Registers Trace registers that are selected by TCBCONTROLB include

TCBCONFIG: The TCB configuration register holds information about the hardware configuration of the TCB.

TCBTW: The trace word read register holds the trace word just read from online trace memory. The TW read is pointed to by the TCBRDP register, which then increments to the next TW in the on-chip trace memory. If TCBRDP is at the maximum size of the on-chip trace memory, the increment wraps back.

TCBRDP: The trace word read pointer indicates the location in the online trace memory where the next trace word will be read. And post-incrementing TW register value to point to the next location. (A maximum value wraps the address around to the beginning of the trace memory.) This is required only for on-chip memory trace.

TCBWRP: The trace word write pointer indicates the location in the online trace memory where the next new trace word will be written. This is required only for on-chip memory trace.

TCBSTP: The trace word read pointer indicates the location of the oldest TW in the online trace memory. This register points to the on-chip trace memory address at which the oldest TW is located. If a continuous trace to on-chip memory wraps around the on-chip memory, TSBSTP will have the same value as TCBWRP. This is required only for on-chip memory trace.

TCBTRIGx: The trigger control registers 0–7 are used to specify some conditions that cause the firing of triggers, and to control the resulting action. Eight trigger control registers are defined. Each register is named TCBTRIGx, where x is a single-digit number from 0 to 7 (TCBTRIG0 is Reg 16). The actual number of trigger registers implemented is defined in the TCBCONFIGTRIG field. An unimplemented register will read all zeros and writes are ignored. Each trigger control register controls when an associated trigger is fired and the resulting action.

12.4   TCB Trigger Logic Overview

The TCB can be configured to include a trigger unit. Most of the actual implementation and functionality are implementation-dependent, but if implemented the base-line structure must be as defined in this section.

Two or more triggers can fire simultaneously. The resulting behavior depends on the trigger action set for each of them and whether they should produce a TF6 trace information output. There are two groups of trigger actions: Prioritized and ORed.

Prioritized Trigger Actions: For prioritized simultaneous trigger actions, the trigger control register that has the lowest number takes precedence over the higher-numbered TCBTRIGx registers. The oldest trigger takes precedence overall.

ORed Trigger Actions: The final trigger is created by ORing of local and an chip probe trigger signals based on the TCBTRIG*x* register action logic outputs. The trigger logic is functionally split in three parts:

- *Trigger Source Logic:* may have a number of source events – chip trigger out, probe trigger, debug mode (DM) indication from the processor core – that can be defined, which cause a trigger to fire when the corresponding source condition is satisfied.
- *Trigger Control Logic:* Eight possible trigger control registers (TCBTRIG*x*, $x = \{0..7\}$) are defined. Each of these registers controls a trigger fire mechanism. They can have each of the trigger sources as the trigger event and they can fire one or more of the trigger actions. This is defined in the trigger control register TCBTRIG*x*.
- *Trigger Action Logic Actions:* Data in TF6 trace format and chip trigger and probe trigger outputs are combined and placed into trace memory.

Two sets of trigger inputs/outputs are defined on the TCB. One set is defined to be chip-internal, and the other set is defined to be part of the probe interface.

TCB trigger input and output

ChipTrigIn – on-chip rising-edge trigger input.

ChipTrigOut – N single-cycle (relative to core clock) high strobe trigger output to an on-chip unit.

TR_TRIGIN – rising probe trigger input.

TR_TRIGOUT – Single-cycle (relative to probe clock) high strobe trigger to be the probe's trigger output.

12.5 PDtrace External Interface

The TCB receives data from the PDtrace™ interface; the processor core is the source. Several control and configuration signals exist on the PDtrace IF.

The TCB registers are accessed through the EJTAG TAP interface on the core. Because the core already implements an EJTAG TAP controller, there is no need to duplicate the entire state-machine in the TCB. The TCB interface uses the (E)JTAG TAP state machine, which are identical to the TJAG TAP state machine. The Trace Control Block TAP Interface Signals (shown below) are based on the output of the state machine and include a serial interface that is synchronous with the EJTAG interface.

- ETT_TCK from the EJTAG TAP controller clock is not an output from the core, but is the input to the TAP controller in the core and is also used by the TCB.
- ETT_TDI is the TDI signal from the EJTAG probe; the TCB must use the same input as the TAP controller in the core.
- ETT_TRST_N is an asynchronous TAP reset from the EJTAG probe.
- ETT_SoftReset in the TAP controller state machine is in the testlogic reset state.

- ETT_Capture is when the TAP controller state machine is in the data-capture state. This indicates that the ETT_Inst[4:0] input is valid.
- ETT_Shift in the TAP controller state-machine is in the data-shift state.
- ETT_Update is when the TAP controller state-machine is in the data-update state. This indicates that the ETT_Inst[4:0] input is valid.
- ETT_Inst[4:0] is the current value of the instruction register in the TAP controller. This selects which TCB register is the target in the capture and update cycles. The options are TCBCONTROLA, TCBCONTROLB, and TCBDATA.
- ETT_TCBData out is the serial output data, synchronous to ETT_TCK's rising edge. When the ETT_Shift is asserted and ETT_Inst[4:0] selects one of the three EJTAG TCB registers, this output must present data.

The timing diagram in Fig. 12.14 shows an access to the TCBCONTROLA register.

In the first two cycles ETT_TRST_N is released, and the selection of an instruction register is started in the TAP controller state machine using ETT_TMS (not used by the TCB TAP). In the first multicycle block, the core TAP controller has its internal instruction register set to 0×10 (= TCBCONTROLA register). This is reflected on ETT_Inst[4:0]. After the other multicycle block, the core TAP controller is in the capture data register state. This is reflected on ETT_Capture. When ETT_Capture is set, the next rising edge on ETT_TCK should update the TCB TAP shift register, with the value of the register selected by ETT_Inst[4:0] (in this case TCBCONTROLA). In the following 32 clock cycles the shift register should receive write data on ETT_TDI and present read data on ETT_TCBData (LSB first on both buses).

One or more cycles after ETT_Shift is de-asserted, the ETT_Update signal will be asserted for one cycle. Assertion of ETT_Update is the signal to write the current contents of the shift register to the register selected by ETT_Inst[4:0] (in this case TCBCONTROLA).

The EJTAG TAP controller will be moved to access other registers, which eventually changes the contents of the ETT_Inst[4:0] pins. Even though ETT_Inst[4:0] is asserted long before ETT_Capture and de-asserted long after ETT_Update, the TCB TAP should only sample the value when either ETT_Capture or ETT_Update is asserted.

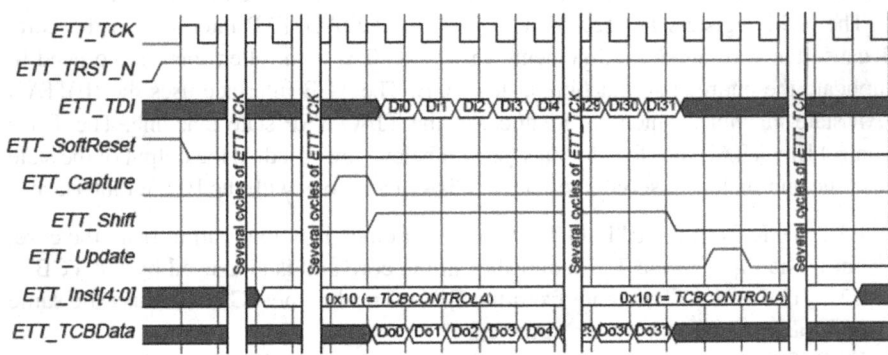

Fig. 12.14 TCB TAP register access timing diagram. *Source*: MIPS Technologies, Inc. All rights reserved

12.6   TCtrace IF

When the TCB is implemented with the ability to send the trace information to a probe, this is done through an intermediate interface called the TCtrace IF. The TCtrace IF is used to connect a small probe interface block to the TCB. The PIB module is the module driving the actual probe I/O pads, which creates the probe IF. The PIB is left as a separate unit, in order to be placed physically near the pads for improved I/O timing. Also the PIB can be more or less advanced with the internal clock-multiplier to enable higher trace bandwidth on a limited number of TR_ DATA trace pins.

The entire TCtrace IF is required in the TCB if off-chip trace memory is implemented; otherwise it is optional. The chip-level trigger input and outputs (ChipTrigIn and ChipTrigOut) are required if one or more trigger control registers are implemented.

The Probe IF can be implemented in a number of widths, allowing a trade-off between the number of pins used and the available bandwidth for tracing. The ratio of the frequency on this interface to the processor core clock frequency can also be configured, to give the maximum possible bandwidth. The PIB module provides the on-chip link between the TCtrace IF and the probe IF. And performed clock multiplication/clock division operations to align the TCtrace and the external interface.

TCtrace Signal Description

- TR_CLK output: Clock to the probe containing the external trace memory. This may be a double-data-rate (DDR) clock, and therefore both of its edges may be significant.
- TR_DATA[15:0] output: Data signals to external trace memory. These may be limited to widths of 4, 8, and 16.
- TR_TRIGIN In yrigger input: Rising-edge trigger input.
- TR_TRIGOUT out trigger output:– Single-cycle trigger output.
- TR_PROBE_N active low input: Indicates that a probe is attached to the device. If this signal is inactive (high), the TR_ outputs can be disabled. It can also be used to control EJTAG signal routing if useful. This signal is optional on a PDtrace-compatible device, but is required on all probes.
- TR_DM output debug mode: When asserted, this indicates that the core has entered debug mode. In a multicore chip, this output can be an AND or an OR or some other function of all the debug-mode indications from each core as specified in the multicore chip documentation.

With on-chip trace memory, the TCB can work in three possible modes:

1. Trace-From mode: In Trace-From mode, tracing begins when the processor enters a processor mode/ASID value THAT is defined to be traced or when an EJTAG hardware breakpoint trace trigger turns on tracing. Trace collection is stopped when the buffer is full, setting a bit in TCBCONTROLB. External probe software, on polling this register, can then read out the internal trace memory.

Saving the trace into the internal buffer will re-commence only when the TCBCONTROLB bit is reset and if the core is sending valid trace data.

2. Trace-To mode: In Trace-To mode, the TCB keeps writing into the internal trace memory, wrapping over and overwriting the oldest information, until the processor reaches an end-of-trace condition. End-of-trace is reached by leaving the processor mode/ASID value, which is trace, or when an EJTAG hardware breakpoint trace trigger turns tracing off. At this point, the on-chip trace buffer is dumped out.

3. Under trigger unit control: If one or more trigger control registers (TCBTRIGx) are implemented and they are using start, end, or center triggers, then the trace mode should be set to Trace-To mode.

12.7 PDTRACE Operations

PDTrace allows four levels of operation:

- No PDtrace implemented.
- PDtrace with PC tracing only.
- PDtrace with PC and load and store address tracing only.
- PDtrace with PC, load, and store address, and load and store data tracing.

Within each level, all features required to support the level must be implemented.

PC Tracing and Trace Compression techniques are used when tracing different values. The methods used differ for each "type" of value being traced. For example, the PC may be sent as a delta from the previous PC address. Sometimes the full PC value needs to be sent when the trace process starts at the beginning of tracing or after a buffer overflow, or for synchronization. In this case, the PC can be sent uncompressed, or some method such as bit-block compression can be used. The PDO_TMode signal allows compressed and full trace modes to be selected for information being currently traced.

When PDO_TMode is zero, the delta of the PC value is transmitted. The PC delta is a signed 16-bit (positive or negative) value computed from the PC value of the instruction executed just before the branch target instruction (the instruction executed in the branch delay slot after a branch instruction).

PC_delta = (new_PC - last_PC)

If the width of the computed delta value is bigger than the width of the PDO_ AD bus, then the lower bits are sent first, followed by the upper bits. When the PDO_TMode value is one, this implies that the full PC value is transmitted. Depending on the width of the bus, this may take multiple cycles.

Load or Store Address Trace and Compression: With a PDO_TMode zero value, the load address transmitted is a delta from the last transmitted load address. Stores are similar, where the computed delta is from the last transmitted store address.

Note that the last load instruction can be a load instruction of any type. The same is true for stores.

load_address_delta = current_load_address - last_load_address

store_address_delta = current_store_address - last_store_address

The delta transmission is effective when the load or store addresses increase or decrease sequentially.

With a PDO_TMode value of one, the value transmitted is the full address of either the load or the store. Depending on the width of the trace bus and the processor data width, this may take multiple cycles to transmit.

Load or Store Data Tracing: Data values have less compression flexibility than instruction information. Compression techniques such as delta values and bit-block compression have been shown to be useful in achieving some compression ratio; however PDtrace does not dictate any compression for data values, and therefore both PDO_TMode values transmit the full data values.

Chapter 13
ARM ETM

ARM embedded trace macrocell (ETM) is a dedicated trace instrument for ARM processors. Like the MIPS EJTAG discussed in the previous chapter, ETM allows the program flow to be passively monitored, along with data and address buses, and to generate a sequencal flow of executed instructions, optionally including with the data accessed. The ETM trace hardware is tightly coupled to the microcontroller core, keeping track of instruction that are executed, and depending on the instruction flow, exporting either a full or compressed (using Branch Trace Messaging) version of the trace format. ARM has a lineage of ETM solutions for its different architecture families. As was discussed with EJTAG, features and interfaces tend to change with evolving families of processors and with their changing debug requirements. For simplicity, we limit our discussion to ETM9, which is debug instrumentation closely associated with the ARM9 family of processors (Fig. 13.1).

An ETM9 enables instructions and data to be traced. The ARM9 core supplies the ETM module with the signals needed to carry out the trace functions. The ETM9 module is operated by means of the trace or JTAG interface. The trace information is stored in an internal FIFO and forwarded to the debugger via the interface. The following trace modes are supported:

- Normal mode with 4- or 8-data-bit width.
- Transmission mode.

    - Full-rate mode at core frequencies <100 MHz.
    - Half-rate mode at core frequencies >100 MHz.
    - Quarter-rate mode at higher core frequencies.

13.1 ETM Signals

An ETM trace port interfaces to all of the signals provided by the ARM ETM and the JTAG run control signals. The ETM trace port and TJAG signals are shown in Fig. 13.1. The signals are summarily described in the following. Section.

N. Stollon, *On-Chip Instrumentation: Design and Debug for Systems on Chip*,
DOI 10.1007/978-1-4419-7563-8_13, © Springer Science+Business Media, LLC 2011

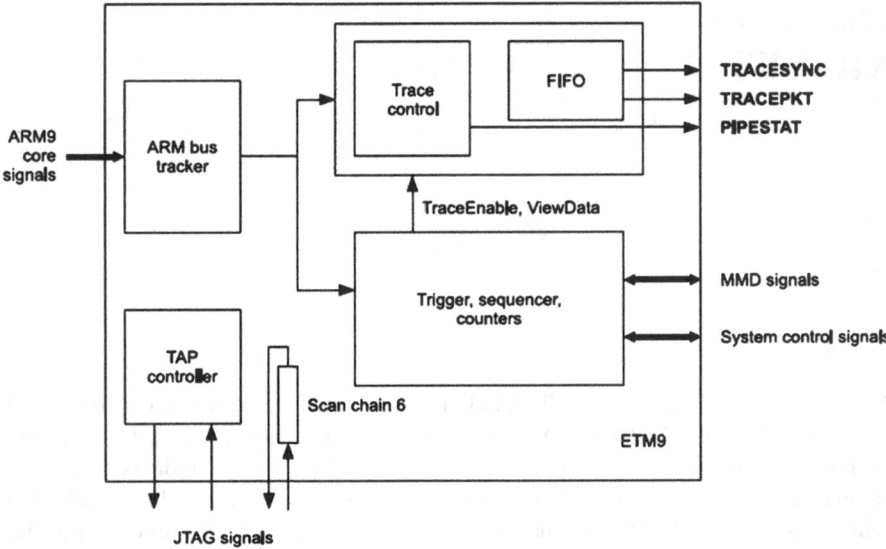

Fig. 13.1 ETM9 module and its interfaces. *Source*: ARM Holdings plc. All rights reserved

Like the other trace formats discussed, the ETM trace is transmitted in packet format that can be configured for export into data buses of varying widths or via a serial (i.e. JTAG) interface.

13.1.1 External Signals

TRACECLK: The trace clock signal provides the clock for the trace port. PIPESTAT[2:0], TRACESYNC, and TRACEPKT[n-1:0] signals are referenced to the rising edge of the trace clock.

PIPESTAT[2:0]: The pipeline status signals provide a cycle-by-cycle indication of what is happening in the execution stage of the processor pipeline.

TRACESYNC: The trace sync signal is used to indicate the first packet of a group of trace packets, and is asserted HIGH only for the first packet of any branch address.

TRACEPKT[n-1:0]: The trace packet signals are used to output packaged address and data information related to the pipeline status. All packets are eight bits in length, irrespective of the number of trace packet signals implemented. There are three cases to consider for how trace packets are output on the trace packet signals:

- 4-bit TRACEPKT bus (TRACEPKT[3:0] signals). A packet is output over two cycles. In the first cycle, Packet[3:0] is output and in the second cycle, Packet[7:4] is output to trace port analyzer or analysis probe.
- 8-bit TRACEPKT bus (TRACEPKT[7:0] signals). A packet is output in a single cycle.
- 16-bit TRACEPKT bus (TRACEPKT[15:0] signals). Up to two packets can be output per cycle. If there is only one valid packet, it is output on TRACEPKT[7:0]. If there are two packets to output, the first is output on TRACEPKT[7:0] and the second on TRACEPKT[15:8].

EXTTRIG: EXTTRIG is an optional signal, intended to be an input to one of the external inputs on the ETM.

DBGRQ: The DBGRQ signal is used by the JTAG interface unit as a debug request signal to the target processor. The DBGRQ signal can be used to enter debug mode after receiving a "BREAK-IN" signal from the logic analyzer through run control. This allows a logic analyzer triggering capability to be used for complex breakpoints.

DBGACK: The DBGACK signal is used to detect entry or exit from the debug state.

Figure 13.2 shows the structure of the TAP interface and its relationship with ETM registers.

The ETM registers are programmed via the JTAG interface into a 40-bit shift register comprising:

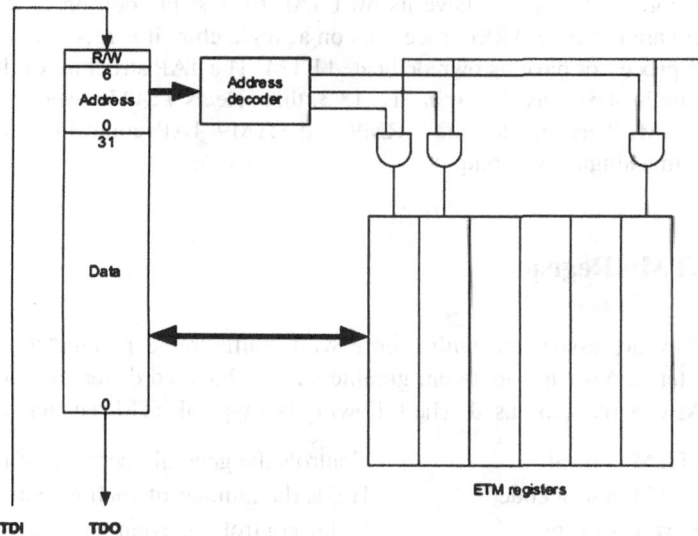

Fig. 13.2 Structure of the ETM TAP. *Source*: ARM Holdings plc. All rights reserved

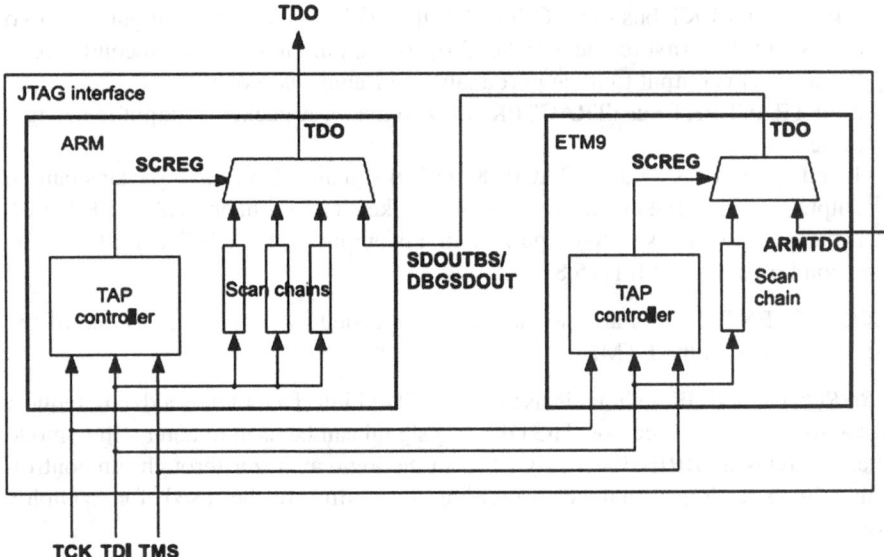

Fig. 13.3 TAP interface between an ARM core and ETM. *Source*: ARM Holdings plc.

- a 32-bit data field
- a 7-bit address field
- a read/write bit.

The ARM core will typically have its own TAP for test in addition to the ETM. Where there are multiple ARM processors on a single chip, it is recommended that each ARM processor have its own dedicated ETM. The TAP structure of the ARM includes a multiplexor, as shown in Fig. 13.3, that selects TDO between the ARM core and ETM. This enables the ARM9 and ETM9 TAP controllers to run in parallel, with a single TDO output.

13.2 ETM9 Registers

Because they are associated with a core with configurable parameters, specific ETM registers are to an extent configurable and are handled differently according to the ETM version being used. The following is a typical ETM register set:

00000000 ETM control	Controls the general operation of the ETM
00000001 ETM config code	Holds the number of each resource
00000010 Trigger event	Holds controlling event
00000011 MMD control	Configures the map decoder
00000100 ETM status	Holds pending overflow status bit
00001000 TraceEnable event	Holds enabling event

00001001 TraceEnable region	Holds include/exclude region
00001010 FifoFull region	Holds include/exclude region
00001011 FifoFull level	Holds the level below which the FIFO is considered full
00001100 ViewData event	Holds the enabling event
00001101 ViewData control 1	Holds include/exclude region
00001110 ViewData control 2	Holds include/exclude region
00001111 ViewData control 3	Holds include/exclude region
0001xxxx Addr. comparator 1–16	Holds the address of comparison
0010xxxx Addr. access type 1–16	Holds the type of access
0011xxxx Data compare values	Holds the data to be compared
0100xxxx Data compare masks	Holds the mask for the data access
010100xx Initial counter value 1–4	Holds initial value of the counter
010101xx Counter enable 1–4	Holds counter clock enable/event
010110xx Counter reload 1–4	Holds counter reload event
0101 11xx Counter value 1–4	Holds current counter value
0110 0xxx Sequencer state/ctrl	Holds the next state triggering events
0110 10xx External output 1–4	Holds controlling event for each output
0111 0xxx Implementation specific	

The ETM9 is a configurable block that can be instantiated with differing amounts of event, trigger, and supporting logic to create trace functions with different complexity levels; the "medium" complexity version of the ETM9 provides the following features:

- Four address comparator pairs.
- Two data comparators with filter function.
- Four direct trigger inputs.
- One trigger output.
- Eight memory-map decoders for decoding the physical address area.
- One sequencer.
- Two counters.

The ETM includes a memory map decode (MMD) block which, to simplify the access of other logic that the core is attached to, decodes address maps using device-specific logic. This logic drives the MMDIN inputs to the ETM, for use in triggering and analysis, in a similar manner to the address comparator and address range comparator resources. The eight MMD regions are decoded in hardware to correspond to different memory regions, as illustrated by the following example:

Address range affected accesses

0000 0000H – 0000 0FFFH instruction cache (I-cache)
0000 1000H – 0000 1FFFH data cache (D-TCM)

0000 2000H – 0FFF FFFFH user RAM access
1000 0000H – 100F FFFFH FLASH access
1010 0000H – 101F FFFFH communication RAM access
2000 0000H – 2FFF FFFFH SDRAM access
3100 0000H – 31FF FFFFH peripheral access
8000 0000H – FFFF FFFFH PCI access

13.3   Trace Interface

In order to read out the trace information collected by the ETM9, a trace port is
used to trace internal processor states. The trace port is controlled, enabled, and
disabled using an external debug probe connected to the JTAG interface. This trace
port uses the TRACEPKT as a trace data port. These TRACEPKT signals are typically multiplexed with other pin signals to preserve pin resources; in many cases
they are shared with the GPIO port. The trace interface can be configured to output
data at a data width supported by the trace port, so for a 16-bit port, trace may be
4, 8, or 16 bits wide. Smaller trace ports are naturally more limited. For example,
if a data width of 4 bits is selected, the TRACEPKT (3:0) signals at GPIO (11:8)
are automatically switched to the trace function. If a data width of 8 bits is assigned,
the TRACEPKT (7:4) signals at GPIO (21:18) are also switched to the trace
function.

Chapter 14
Infineon Multicore Debug Solution

Infineon MCDS is a multicore debug solution developed for their own chips. Like other debug solutions discussed, it consists of configurable IP building blocks, which provide trace compression, trace qualification, timestamping, and complex cross-target triggering. It also enables measurement of several performance indicators in parallel with timestamped trace results.

Figure 14.1 shows the MCDS sub-system consisting of a MCDS kernel and on-chip trace memory (TMEM). In this example, communication between the on-chip debug environment and the debug tool is implemented based on a JTAG TAP with optional data trace interfaces. Infineon was a long time member of Nexus Forum and its interfaces are in areas partially compliant with Nexus ports (see Chap. 11). Each debug target (processor core, bus) is connected to the MCDS through an adaptation logic block. The design of such a block may be target-specific. Each block adapts the target's custom interface to a generic standardized interface that is used by MCDS. It also synchronizes signals from the target side to the clock domain of the MCDS in case they are in different clock domains.

The architecture of the MCDS kernel depends on the number and type of debug targets and consists of so-called observation blocks (OB), a multicore cross connect (MCX), and a debug memory controller (DMC). The MCX is connected to all OBs and the DMC. It is responsible for distribution of cross triggers, which are programmable, and provides a central timestamp for all trace messages. Additionally, MCX provides a number of counters, which can be used to count events and trigger an action after an event has occurred n times or a certain time period has elapsed. MCX provides the functionality to observe a system with multiple processor cores, where interactions between the cores take place and complex conditions have to be evaluated to recognize a certain event.

Each target signal within the SoC is connected to its dedicated OB. Within this block, trace qualification and trace message generation take place. Each OB may contain several custom trace units of different types.

To start or stop the trace recording, generate cross triggers, and control the targets, trace qualification logic is implemented as shown in Fig. 14.2. The trigger logic, based a AND/OR matrix operating on direct or negated inputs, and triggering on edge or level logic conistions is implemented for all trace qualification blocks contained in the OBs and in the MCX.

N. Stollon, *On-Chip Instrumentation: Design and Debug for Systems on Chip*,
DOI 10.1007/978-1-4419-7563-8_14, © Springer Science+Business Media, LLC 2011

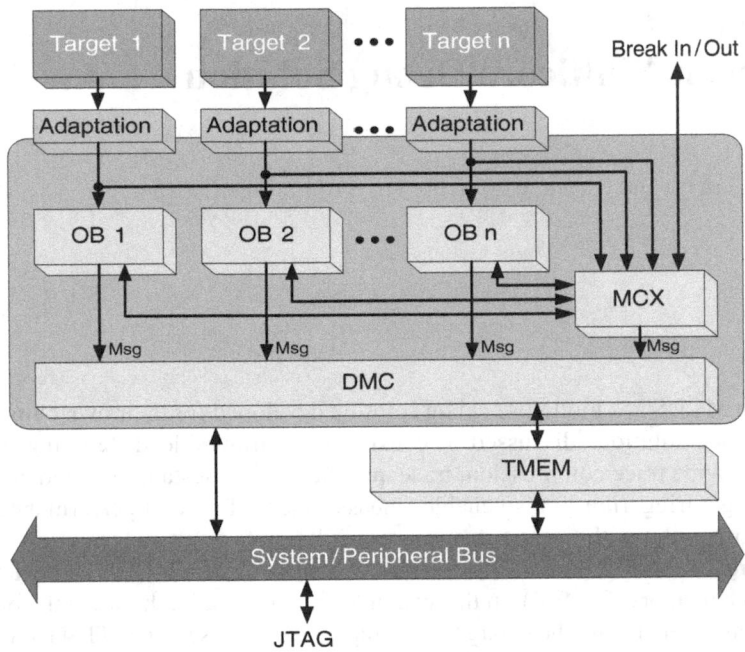

Fig. 14.1 MCDS sub-system. *Source*: Infineon Technologies AG. All rights reserved

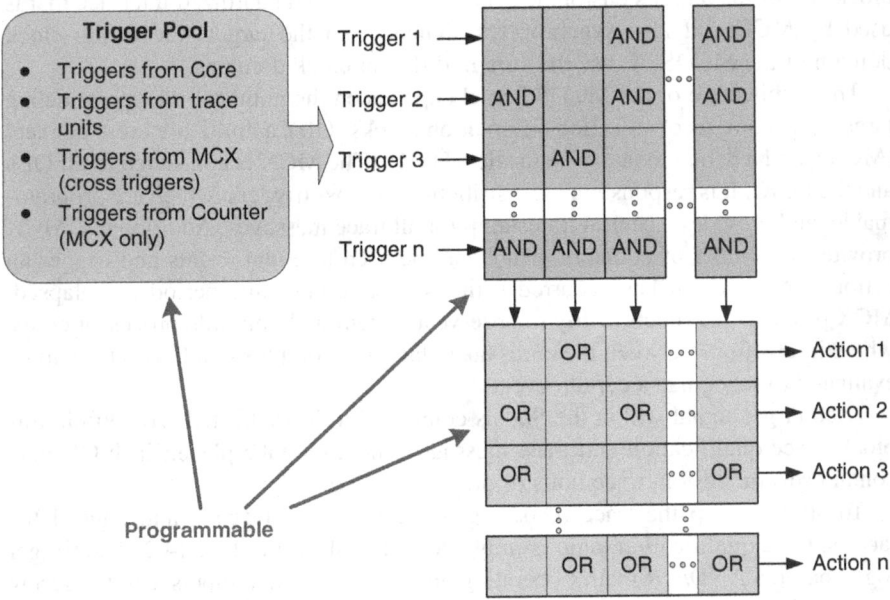

Fig. 14.2 MCDS trace qualification. *Source*: Infineon Technologies AG. All rights reserved

From the perspective of the debug tool, the MCDS has to be programmed for a certain debug or trace task by writing configuration information into a set of memory-mapped registers. These registers control the AND/OR matrices of each OB, the DMC, and a number of trigger sources (e.g., address/data comparators).

14.1 MCDS Trace Protocol Definition

The basic interface is a synchronous tagged data protocol without handshake. The sender places the data in well-defined packets on the data port and indicates concurrently on the mode port which kind of packet is present.

Merging the mode into the data packet is avoided to simplify the implementation. The mode port must be able to express at least two different values: IDLE and VALID. If different kinds of data are supported, VALID is replaced by other mode port values as shown in Table 14.1.

The MODE port can be used to propagate a response to the debugger via a bus observer block between this trace interface and the system bus (Table 14.1).

If a time-out mechanism is provided, a synthetic FORGET is used. Protocol errors are also forwarded downstream as FORGET. As no back channel is provided, the offending message is dropped by the receiver. Some fine points about the protocol implementation are as follows:

- MCDS processor cores. The smallest common denominator is the interface described here, comprising of two ports:

  - Base address: After power on and after each discontinuity of the program flow, the trace logic needs to know the exact and complete value of the instruction pointer.

Table 14.1 MCDS interface mode encoding

Mode	Description
IDLE	Do nothing, the data port holds no value. The data previously transferred is still valid.
VALID	A new value of the implicitly defined (only possible) format of the receiver is on the data port.
BYTE	A new value of the given format is on the data port.
HALFWORD	
WORD	
DOUBLE	
READ	A new transaction of the given direction is on the data port.
WRITE	
FORGET	The data previously transferred is no longer valid. The data port may hold additional information during the first clock cycle (see Program Trace below).

Source: Infineon Technologies AG. All rights reserved

- Instruction pointer increment: Once the base is known, only the incremental updates required to keep the local copy in sync with the original instruction pointer.

- Discontinuities are either:

  - Direct: Branches are caused by jump instructions in the executed program. The target address is a constant (label) in the source code and can be obtained from there by the decoder software. A branch of this kind is indicated by FORGET (see Table 14.1) on the increment port.
  - Indirect: Branches are caused by jump instructions with calculated target address (e.g., return from subroutine) or by exceptions (e.g., interrupts, traps). In these cases, the target address must be contained in the trace memory. FORGET on the base port is used to indicate such a branch. Each FORGET received on any of the two ports invalidates the current base address. The exact protocol definition is given in Table 14.2.

There is no need for the sender to serve both ports concurrently. The only requirement is that the base is sent prior to (or at least concurrently with) the next discontinuity. If this is not possible, the sender may set both mode ports to FORGET, as this is interpreted as overrun (Fig. 14.3).

After each branch, the instruction pointer is unknown to the trace logic until a new base address is received. However, the decoder software may already know the address (e.g., L1 in Fig. 14) from the source code.

Table 14.2 Discontinuity protocol

	Base Address Protocol
Mode Port	*Data Port*
IDLE	Don't care.
VALID	Target address after a preceding discontinuity, optionally the current instruction pointer otherwise.
FORGET	Target address after a preceding discontinuity. This base must only be used for one clock cycle and discarded thereafter. Don't care otherwise.
	Instruction Pointer Increment Protocol
Mode Port	*Data Port*
IDLE	Don't care.
VALID	Instruction pointer increment. This is the number of bytes the instruction pointer was advanced since the last time the Mode Port was not IDLE.
FORGET	Instruction pointer increment. This is the number of bytes the instruction pointer was advanced linearly since the last time the Mode Port was not IDLE. In case of a taken branch this includes the branch instruction.

Source: Infineon Technologies AG. All rights reserved

		base port	Incr. port		Reconstructed IP
L0:		VALID L0	IDLE		L0
	<instructions>	IDLE	VALID d1		L0 + d1
	<instructions>	IDLE	VALID d2		L0 + d1 + d2
		IDLE	IDLE		
	<instructions>	IDLE	VALID d3		L0 + d1 + d2 + d3
	JMPA L1	IDLE	FORGET d4		L0 + d1 + d2 + d3 + d4 + direct branch taken
		IDLE	IDLE		<undefined>
L1:	<instructions>	VALID L1	VALID d5		L1 + d5
	<instructions>	IDLE	VALID d6		L1 + d5 + d6
	JMPI L2	FORGET	VALID d7		L1 + d5 + d6 + d7 + indirect branch taken
		IDLE	IDLE		<undefined>
L2:	<instructions>	VALID L2	VALID d8		L2 + d8
	<instructions>	IDLE	VALID d9		L2 + d8 + d9
	<instructions>	IDLE	VALID d10		L2 + d8 + d9 + d10

TRACE_EXAMPLE

Fig. 14.3 Trace for minimized messages. *Source*: Infineon Technologies AG.

For multiscalar processors, the last increment leading to a taken branch (e.g., d4 in Fig. 14.3) may include more than the branch itself. It is therefore not guaranteed that the branch instruction is stored at address L0+d1+d2+d3.

In the case of some exceptions (e.g., illegal target address) the target address must be analyzed to distinguish the exception from a taken branch. That is why it is important to treat exceptions and interrupts as indirect branches.

14.1.1 Data Trace

To trace transactions on an arbitrary bus system, consisting of address, data, and control information, the following are needed:

- The effective address (byte granularity).
- The current data (size depending on transaction).
- Auxiliary information (bus mode) such as mastership and privileges.

The mode of the third item is used to signal completeness. Whenever asserted, READ or WRITE information concurrently valid is considered to belong to the same transaction. The mode port of transaction type is READ or WRITE for exactly one clock cycle for each transaction (Table 14.3).

Ownership is used to refer to a task ID or process ID. The generic OTU is able to process the ownership information of an arbitrary processor core if implemented in hardware. As the rate of change for the process ID is rather low, it will often be sent multiplexed over other trace interface signal lines of the core. This is legal, provided a dedicated signal to drive the mode port is available. If the core is not doing any useful work (e.g., if no task is active), the process ID should be invalidated by FORGET.

Table 14.3 Transaction protocol

Transaction Address Protocol	
Mode Port	*Data Port*
IDLE	Don't care.
VALID	Address (byte accurate) sent by the master to the slave.
FORGET	Don't care.
Transaction Data Protocol	
Mode Port	*Data Port*
IDLE	Don't care.
BYTE	Data byte (8 bit, right justified) written by the master or read from the slave.
HALFWORD	Data half-word (16 bit, right justified) written by the master or read from the slave.
WORD	Data word (32 bit, right justified) written by the master or read from the slave.
MIS48	Misaligned double-word (48 bit, right justified) written by the master or read from the slave.
DOUBLE	Data double-word (64 bit, right justified) written by the master or read from the slave.
FORGET	Don't care
Transaction Type Protocol	
Mode Port	*Data Port*
IDLE	Don't care.
WRITE	Additional information (e.g. master ID, privilege level).
READ	Additional information (e.g. master ID, privilege level).
FORGET	Don't care.

14.2 Debug Transactor: RUN Control Bus Master

A basic bus transactor implementation supports simple read and posted write data operations and may require stalling between operations to ensure synchronization of signaling more advanced operations. More advanced operations such as bursting may require additional (dedicated) logic (Fig. 14.4).

A transactor bus master operation can be initiated from either an external register load or from trigger output signals acting on specific bus-monitoring operations. Address and data for individual bus transactions can also be written from the externally controlled registers, although this may be a slower manual process or require multiple cycles. Alternately, writing of regular (i.e., incrementing or other simple pattern) address and data can be controlled by counters or by logic enabled by trigger signals. Data signals or performance data may also be traced (i.e., sequentially or periodically) with the bus master operations optionally stalled during the JTAG data-download phase, avoiding loss of continuity. Additional trigger or state signals may be used for defining and controlling basic memory maps or domains.

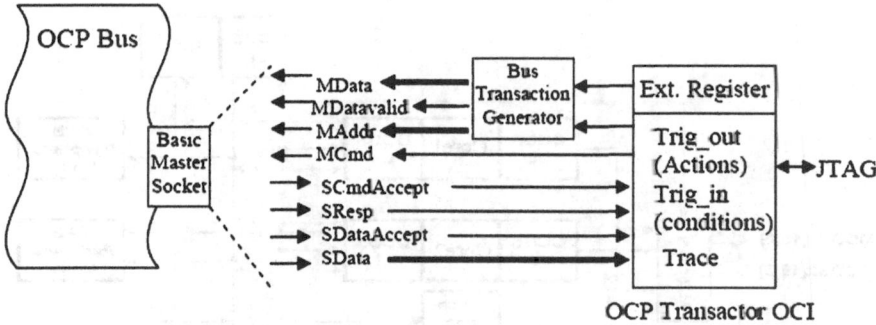

Fig. 14.4 A bus master transactor

14.3 MCDS Run Control: On-Chip Debug Support

MCDS supports three levels of debug operation:

Level 1 is for use for real-time software debugging operations based on a JTAG interface that is used by the external debug hardware to communicate with the system. The on-chip Cerberus bus master module controls the interactions between the JTAG interface and the on-chip modules. The external debug hardware may become master of the internal buses and may read or write to the on-chip register/ memory resources. The Cerberus also makes it possible to define breakpoint and trigger conditions, as well as to control user program execution (run/stop, break, single step).

Level 2 makes it possible to implement program tracing capabilities for enhanced debuggers by extending the level 1 debug functionality with an additional 16-bit-wide trace output port with trace clock. With the trace extension, trace capabilities are provided for several cores and IP-blocks with just one trace being active at a time.

MCDS level 3 is based on a multicore debug solution using a special emulation device that has additional features required for high-end emulation purposes. It does not use more interface signals, but replicates the debug interface for many cores and pro-vides two out of *N* simultaneous trace channels differentiated by the process ID port.

Components in Fig. 14.5 include:

- Cerberus OCDS system control unit (OSCU).
- Cerberus multicore break switch (MCBS) (cross-trigger unit with extensions).
- Cerberus JTAG debug interface (JDI).
- Suspend functionality of the peripherals (stop block activity for debug purposes).
- Several level 1 and level 2 units for the cores and IP blocks.
- BCU that allows cross-triggering by the system bus events.

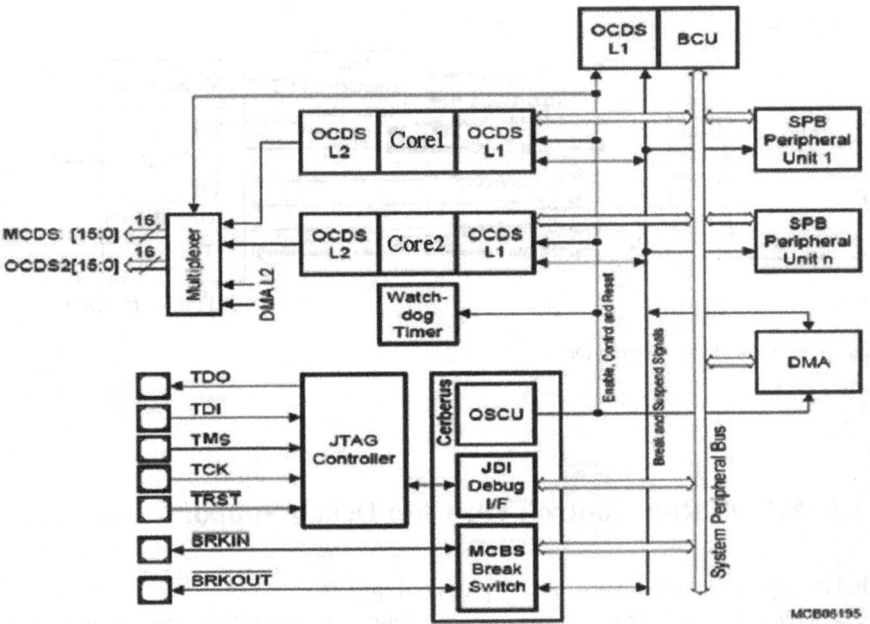

Fig. 14.5 Debug system block diagram. *Source*: Infineon Technologies AG. All rights reserved

The main philosophy of the cores is that the complete architecture and the status of a target system are visible from its memory-map address space, including on-chip memories, processor core registers, and the register of the peripheral units.

A typical level 1 debugging configuration includes:

1. The debugger software, supporting a standard JTAG protocol via a PC port.
2. The debugger hardware adapter, connecting the JTAG interface.

The processor core provides Cereberus with the following two basic parts:

- Debug event trigger generation.
- Debug event trigger processing.

The first part controls the generation of debug events and the second part controls what the actions are from the triggers.

Debug Event Generation: If debug mode is enabled, debug events can be generated by:

- Debug event generation from debug triggers.
- Activation of the external break input pin BRKIN.
- Execution of a DEBUG instruction.
- Execution of an MTCR/MFCR instruction.

Debug Actions: Four types of debug actions are available:

Table 14.4 MCDS debug registers

Register Short Name	Register Long Name	Address
DBGSR	Debug Status Register	F7E1 FD00$_H$
EXEVT	External Break Input Event Specifier Register	F7E1 FD08$_H$
CREVT	Core SFR Access Break Event Specifier Register	F7E1 FD0C$_H$
SWEVT	Software Break Event Specifier Register	F7E1 FD10$_H$
TR0EVT	Trigger Event 0 Register	F7E1 FD20$_H$
TR1EVT	Trigger Event 1 Register	F7E1 FD24$_H$
CPU_SBSRC	CPU Software Break Service Request Control	F7E0 FFBC$_H$[a]

[a]Located in the CPU slave (CPS) interface register area
Source: Infineon Technologies AG. All rights reserved

- Assert BRKOUT signals by the MCBS unit.
- Halt the processor core.
- Cause a breakpoint trap.
- Generate an interrupt request.

These debug actions are selected by programming the corresponding event specifier registers which determine the action taken when the corresponding debug event occurs (Table 14.4).

14.3.1 BCU Level 1 (Bus-Observer Unit on the System Bus)

The BCU on the system bus supports both level 1 and means for breakpoint generation. The BCU contains one comparator for the following:

- The arbitration phase (look for specific bus master).
- The address phase (look for specific address or range).
- The data phase (look for read, write, supervisor mode, etc.).

The results can be combined to generate a break request signal, to be sent to the break switch (cross trigger block).

For a level 2 trace, in every trace clock cycle, 16 bits of core trace information are sent out, representing the current state of the cores. The trace output lines are grouped into three parts:

- 5 bits of pipeline status information.
- 8-bit indirect PC bus information.
- 3 bits of breakpoint qualification information.

With this information, an external debugger can reconstruct a cycle-by-cycle image of the instruction flow. The trace information can be captured by the external debugger hardware and used to rebuild later (off-line, using the source code) a cycle-accurate disassembly of the code that has been executed. It is also possible to follow in real time the current PC, facilitating advanced tools such as profilers and coverage analysis tools.

The trace output port is controlled by the OSCU. The trace data can be output at processor clock speed. The trace clock can be higher if two cores are traced or if a better compression of trace data of all cores can keep the trace clock low.

14.3.2 Concurrent Debugging in Level 3 MCDS (Two-Channel Tracing)

A concurrent debugging is possible when the control port is used as the second channel and the ownership port is extended with process ID to differentiate between two sources that are traced. The debug setup must define which two cores or IP blocks were selected for concurrent tracing.

14.3.3 Debug Interface (Cerberus) (Debug Bus-Transactor Module)

The Cerberus module is the on-chip unit that controls all levels' main debug functions. Generally, the Cerberus should not be used by any application software, because this could disturb the emulation tool behavior.

The Cerberus module is built up by three parts:

- OCDS system control unit debug bus master.
- JTAG debug interface JDI.
- Multicore break switch cross-trigger unit.

14.4 RW Mode and Communication Mode

As the name implies, the RW mode is used by a JTAG host to read or write arbitrary memory locations via the JTAG interface. The RW mode needs the FPI bus master interface of the Cerberus to actively request data reads or writes.

In communication mode, the Cerberus has no access to the system bus and communication is established between the external JTAG host and a software monitor (embedded in the application program) via the Cerberus registers. The communication mode is the default mode after reset.

In communication mode, the external JTAG host is master of all transactions and requests the monitor to write or read a value to/from the Cerberus COMDATA register. The difference to RW mode is that the read or write request is not actively executed by the Cerberus, but it sets request bits in the processor-accessible register to signal the monitor that the debugger wants to send.

14.5 Multicore Break Switch

In this example, there are two main processor units, the processor core 1 and the PCP2 (a co-processor) core 2. For debugging purposes, the OCDS run control of one processor unit can break (interrupt) the other processor unit or vice versa. The run control tasks are handled by the MCBS unit, which is part of the Cerberus. Figure 14.6 shows the break signal interfaces of this MCBS unit.

The MCBS unit supports the following features (very similar to the OCP debug standard):

- Two independent break-out master units (Core 1 and Core 2).
- Six break-in sources (processor core, PCP, DMA, SBCU, MLI0, MLI1).
- Two port pins, BRKIN and BRKOUT.
- Two independent break buses (two out of N).
- Suspend generation supports delayed suspend.
- Break-to-suspend converter.
- Create interrupt request with a break coming from a source.
- Synchronous restart of the system.

The MCDS is designed to support complex multicore/debugging environments; several debugger applications may have to share a common resources, which may include registers, trace buffers, and the JTAG interface (Table 14.5).

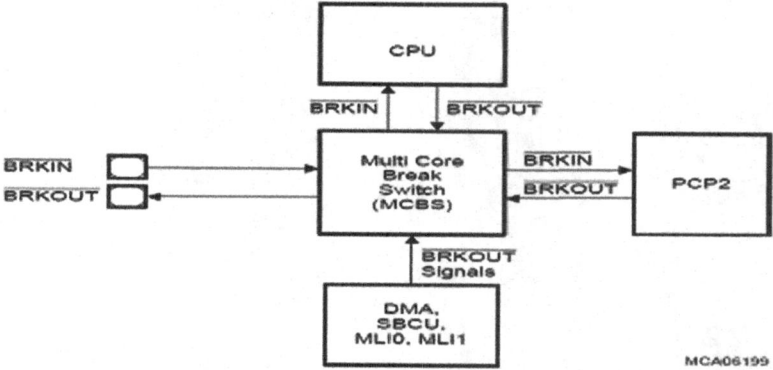

Fig. 14.6 Break switch interfaces. *Source*: Infineon Technologies AG.

Table 14.5 Cerberus bus master registers

Register short name	Register long name	Address
OJCONF	OSCU Configuration by JTAG Register	(a)
CBS_OEC	Cerberus OCDS Enable Control Register	F000 0478$_H$
CBS_OCNTRL	Cerberus OSCU Configuration and Control Register	F000 047C$_{H0}$
CBS_OSTATE	Cerberus OSCU Status Register	F000 0480$_H$
CLIENT_ID	Cerberus JTAG Client Identification Register (32-bit)	(a)
IOCONF	Configuration Register (12-bit)	(a)
IOINFO	State Information for Error Analysis Register (16-bit)	(a)
IOADDR	Address for Data Access Register (32-bit)	(a)
IODATA	RW Mode Data Register (32-bit)	(a)
CBS_JDPID	Cerberus Module Identification Register	F000 0408$_H$
CBS_COMDATA	Cerberus Communication Mode Data Register	F000 0468$_H$
CBS_IOSR	Cerberus Status Register	F000 046C$_H$
CBS_INTMOD	Cerberus Internal Mode Status and Control Register	F000 0484$_H$
CBS_ICTSA	Cerberus Internal Controller Trace Source Address Register	F000 0488$_H$^{0}
CBS_ICTTA	Cerberus Internal Controlled Trace Target Address Register	F000 048C$_H$
CBS_MCDBBS	Cerberus Break Bus Switch Configuration Register	F000 0470$_H$
CBS_MCDBBSS	Cerberus Break Bus Switch Status Register	F000 0490$_H$
CBS_MCDSSG	Cerberus Suspend Signal Generation Status and Control Register	F000 0474$_H$
CBS_MCDSSGC	Cerberus Suspend Signal Generation Configuration Register	F000 0494$_H$
CBS_SRC	Cerberus Service Request Control Register	F000 04FC$_H$

[a]These registers are only accessible via the JTAG interface

Chapter 15
EJTAG and Trace in Toshiba TX Cores

Toshiba supports debug using a version of EJTAG interface revision 1.5, which was released in the late 1990s. This diverges from the current MIPS-EJTAG interface revisions, so a MIPS EJTAG debugger to would not support Toshiba MIPS architecture based parts (and vica versa). Likewise, debug instruction and registers are different from current MIPS. Like other versions of EJTAG, the Toshiba EJTAG interface is an extension to the IEEE 1149.1 JTAG interface. Additional status pins and debug clock signals, in conjunction with JTAG pins, provide real-time PC trace information. Because serial bus access to the memory in the external processor probe is available through the JTAG interface, the debug program can be placed in the external memory. Access to all resources' connected to the processor is available by the DMA function through JTAG interface. The debug support unit (DSU) in the Toshiba MIPS core has an 8-double-word scratch pad memory (MIB), which reduces communication time through JTAG interface.

The following are some of the areas where the Toshiba EJTAG diverges from the more current EJTAG specification:

- Instruction address break.
- Data bus break.
- Processor bus break.
- Hardware debug interrupt.
- Reset, NMI, interrupt mask.
- Instructions for debug – SDBBP, DERET, CTC0, CFC0.
- CP0 registers for debug – Debug, DEPC, DESAVE.

EJTAG interface signals are the main debug connection. Basic debug functions can be used by connecting GTCK, GTRST, GTMS, GTDI, and GTDO to an external processor probe. These are logic equivalent to the standard 1149.1 JTAG interface signals. GTDOE is the output-enable signal for GTDO. GDCLK, GPCST[8:0], and GTPC[3:1] are Toshiba-specific signals used for PC trace.

As a side note: Toshiba has a system bus for MIPS architectures called G-Bus. They put a G prefix on any bus signal that comes close to the core.

During PC trace the GTID and GTDO signals disable their JTAG-related functions and are respectively used for:

N. Stollon, *On-Chip Instrumentation: Design and Debug for Systems on Chip*,
DOI 10.1007/978-1-4419-7563-8_15, © Springer Science+Business Media, LLC 2011

- Debug interrupt of PC trace (GDINT).
- PC output bit 0 of PC trace (GTPC[0]).

The other signals used for PC trace are:

- GTDOE putput-enable signal of test data output.
- GDCLK output clock of PC trace (1/3 processor CLK).
- GPCST [8:0] output status information of PC trace.
- GTPC [3:1] PC output bit [3:1] of PC trace.

The PC trace is driven by the GDCLK, which originates on the chip and can be asynchronous to GTCK.

15.1 Processor Access Overview

The TX core accesses the external processor probe and reads/writes the external monitor memory, registers, and other external resources.

By enabling the processor probe, instructions in the external memory can be executed. Access to the monitor is allowed only when this processor is in debug mode (DM = 1).

The address of the external monitor memory is set in the JTAG_Address_Register by the processor. The data written to or read from an external agent is transferred through the JTAG_Data_Register. JTAG_Control_Register is used to control processor access.

The TX49 core can implement DMA to the internal registers connected to the internal processor bus, host system peripheral, and system memory through the JTAG interface. By using this function, the system memory can be read or written by the external processor probe. The address to perform DMA is set in the JTAG_Address_Register by using the external processor probe. The data written or read by DMA is transferred to the internal processor bus through the JTAG_Data_Register. JTAG_Control_Register is used to control DMA.

The MIB (monitor instruction buffer) is an optional 64-bit (8 double-words) scratch pad memory used to transfer data between the core and external agent on MIB access. The MIB control register is used to control MIB accesses to the processor core.

Parameters used for the monitor program during debug, and parts of the monitor program, can be set in this memory. The monitor program reads and writes the MIB with values from the MIB data register.

There are several types of simple hardware breakpoints defined in the EJTAG specification. These stop the normal operation of the processor and force the system into debug mode. The break occurs when certain activities take place on the processor address, data, and control buses. The debug exception occurs before the bus transaction occurs, preserving any content in the register file or memory. Hardware breaks, unlike software breaks, can be made based on the address on the memory bus, so breakpoints can be set for access to any area of memory. Hardware breaks also enable breaks on load/store operations.

Generally, breakpoints are set up in debug mode and become operational when normal operational mode is re-entered. Using the EJTAG DMA circuitry, the developer can enable breakpoints in normal mode so that the system maintains real-time operation up to the moment of the breakpoint encounter. There are 45 break channels defined in the specification. With the maximum amount of breakpoint hardware, it is possible to have up to 45 concurrent breakpoints set, with each operating independently with separate breakpoint values.

15.2   Toshiba EJTAG Instructions and Registers

EJTAG instructions share the 8-bit IR field between EJTAG instructions for the TX debug support unit, with JTAG codes for standard JTAG instructions (EXTEST, SAMPLE/PRELOAD, INTEST, IDCODE, and HI-Z). EJTAG instructions are as shown in the following:

EJTAG_ImpCode selects the implementation register.
JTAG_ADDRESS_IR selects the JTAG_Address_Register.
JTAG_DATA_IR selects the JTAG_Data_Register.
JTAG_CONTROL_IR selects the JTAG_Control_Register.
JTAG_ALL_IR selects the JTAG_Address_Register, JTAG_Data_Register, and JTAG_Control_Register and serially connects them together into a single value.

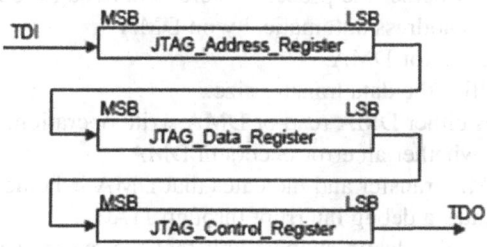

PCTRACE selects the PCTRACE instruction.

0xA0–0xAF MIB_WRITE_DEC selects the MIB data register.
0xB0–0xBF MIB_WRITE_INC selects the MIB data register.
0xC0–0xCF MIB_READ_DEC selects the MIB data register.
0xD0–0xDF MIB_READ_INC selects the MIB data register.
0xE0–0xEF MIB_CONTROL selects the MIB control register.

The EJTAG interface has the following registers:

- *Instruction register*: 8-bit instruction register (required by JTAG).
- *Bypass register*: 1-bit bypass register (required by JTAG).

- *Device identification register* (required by JTAG).
- *Implementation register*: defines the parameters as listed here for usable debug functions for a given implementation:

  o MIPS32/64 – 32- or 63-bit debug function register length.
  o InstBrk – instruction address break available.
  o DataBrk – data bus break available.
  o ProcBrk – processor bus break available.
  o PCSTW – width of PCST output for PC trace.
  o TPCW – width of TPC output for PC trace.
  o NoDMA – DMA available through JTAG.
  o NoPCTrace – PC trace available.
  o MIPS16 – Support for MIPS16.
  o IcacheC – instruction cache coherency available on DMA implementation.
  o DcacheC – data cache coherency available on DMA implementation.
  o PhysAW – physical address length.
  o MIB – monitor instruction buffer availability.

JTAG_Data_Register is used to transfer data between the core and external agent during processor access and DMA implementation.

JTAG_Address_Register is a 36-bit register used to transfer addresses between the core and external agent during processor access and DMA implementation.

JTAG_Control_Register is used to control a variety of EJTAG functions as listed here and to observe the processor state:

- BrkSt – Indicates whether the processor core is in debug mode.
- Dinc – Increments address automatically on DMA.
- Dlock – Locks a bus for DMA.
- Dsz[1:0] – Specifies the data transfer size.
- Drwn – Specifies either DMA read or DMA write operation.
- Derr – Indicates whether an error occurs in DMA.
- Dstrt – Starts DMA transfer and indicates that DMA is being implemented.
- JtagBrk – Generates a debug interrupt through JTAG.
- ProbEn – Informs the debug support unit that the external processor probe is connected.
- PrRst – Resets the processor core.
- DmaAcc – Requests DMA.
- PrAcc – Sets when the processor core issues an access request to an external processor probe; the external processor probe accesses memory in the external system and writes to the JTAG_Data_Register if required, then resets PrAcc.
- PRnW – Indicates whether the processor access is a read or write operation.
- PerRst – Resets the peripheral circuits except the processor core on SoC.
- Run – Indicates whether the processor core is in halt mode.
- Doze – Indicates whether the processor core is in doze mode.
- Sync – Sets whether to start PC trace synchronously to DERET.

- PCLen – Specifies the output length of the target PC of PC trace.
- MibEn – Makes MIB usable.

15.3 Debug Exceptions

A useful element for software debug is a high-priority debug exception (with a higher priority than all other exceptions). The debug exception can occur when a software debug breakpoint instruction is encountered, a single-step instruction occurs, a JtagBrk debug event is registered by the EJTAG circuit, or a hardware breakpoint occurs. When the debug exception occurs, the processor switches into debug mode, where there are no restrictions on access to coprocessors and memory and where the usual exceptions like address error and interrupt are masked.

The debug exception handler is provided by the debug system and can be executed through the EJTAG port using the processor access circuitry, or it can be placed in application code space if it is required.

Exception processing in debug mode (DM bit is set) means that all interrupts including NMI are masked. When the NMI interrupt occurs during debug mode, it is stored internally. The NMI interrupt is taken after the debug handler is finished (DM is cleared).

On debug exception processing, the DEPC and debug registers are updated. The registers other than the DEPC and debug registers retain these values.

The following three types of debug exceptions are supported:

- Debug single-step when the SSt bit in the debug register is set, a debug single step occurs whenever each instruction is executed.
- Debug breakpoint exception occurs when the SDBBP instruction is operated.
- JTAG break exception occurs when the Jtagbrk bit is set in the JTAG_Control_ Register.

During real-time debug operation, both the Debug single-step and Debug breakpoint exceptions are disabled.

15.4 Processor Debug Instructions and CP0 Registers

The following processor instructions and CP0 registers are added for debug:

- SDBBP instruction – software debug breakpoint.
- DERET instruction – debug exception return.
- CTC0 instruction – move control from co-processor 0.
- CFC0 instruction – move control from co-processor 0.

The software debug breakpoint (SDBBP) instruction is defined for the MIPS instruction set architecture and for the code compression application-specific extension MIPS16. For simple breakpoints, the debug system can replace application code instructions with software breakpoint instructions and generate a debug exception.

For leaving the debug mode, a debug exception return (DERET) instruction is also defined. When it occurs, the system leaves the debug mode and normal execution of application and system code resumes.

Debug registers are the DEBUG, DEPC, and DESAVE registers, which are added to the MIPS co-processor 0 (CP0). The DEBUG register shows the cause of the debug exception and any other standard exceptions which may have occurred at the same time. Also, it is used to set up single-step operations. The DEPC or debug exception program-counter register holds the address of where the debug exception occurred. This is used to resume program execution after the debug operation finishes. Finally, the DESAVE or debug exception save register is a scratch pad for one of the general-purpose 32-bit registers of the processor. This frees the general-purpose registers from duty in handling the debug exception handler, which executes without affecting the contents of any of the general-purpose registers.

DEBUG register – Debug configurations and status holds the information for the debug handler. Key values in the debug register include:

- DM debug mode indicates that a debug exception has taken place. This bit is set when a debug exception is taken and is cleared on return from the exception (DERET). While this bit is set, all interrupts, including NMI, TLB exception, BUS error exception, and debug exception, are masked and the cache line-locking function is disabled.
- OES other exception status is set to indicate that an exception other than reset, NMI, or a TLB refill has occurred at the same time as a debug exception.
- SSt is set to 1 to indicate the single-step debug function is enabled.
- DINT (debug interrupt break exception status) is set to 1 when debug interrupts occur.
- DIB (debug instruction break exception status) is set to 1 on instruction address break.
- DDBS (debug data break store exception status) is set to 1 on data address break at store operation.
- DDBL (debug data break load exception status) is set to 1 on data address break at load operation.

DEPC – The debug exception PC register.

DESAVE – The debug SAVE register.

The debug support unit also has ranges of registers used to set up breakpoints. Accessing these registers is allowed only when the processor is in debug mode (DM = 1). In other modes (DM = 0), accessing these registers will cause an address error.

The debug control register is used to control debug functions:

- Select whether to stall the processor and output all bits of PC or abort output of the target PC without stalling the processor.
- Indicate the core is in halt or doze mode when a debug exception occurs.

The following relate to instruction, data, or processor break:

- Instruction address break status register shows the instruction address break status.
- Instruction address break address register 0 is used to specify the instruction breakpoint in the virtual address.
- Instruction address break control register 0 controls an instruction address break, allowing a debug exception by an instruction address break or output of trace trigger by an instruction address break.
- Instruction address break address mask register 0 used to specify the mask bits for comparison of the instruction address breakpoint.
- Data bus break address register 0 specifies the data bus breakpoint in virtual address.
- Data bus break control register 0 controls a data bus break by allowing active debug exceptions by a data bus break, an output of trace trigger by a data bus break, and/or the byte to be masked of the data value to make a break occur.
- Data bus break address mask register 0 is used to specify the mask bits for comparison of the data bus breakpoint address.
- Data bus break value register 0 is used to specify the data bus breakpoint value.
- Processor bus break address register 0 is used to specify the processor bus breakpoint in physical address.
- Data bus break status register shows the data bus break status and number of data bus break channels.
- Processor bus break data register 0 is used to specify the processor bus breakpoint value.
- Processor bus break data mask register 0 is used to specify the mask bits for comparison of the processor bus breakpoint value.
- Processor bus break control register 0 controls a processor bus break by allowing debug exception by a processor bus break, trace trigger by an instruction address break, or on instruction fetch or data read or write, by making a processor bus break occur.
- Processor bus break status register stores the processor bus break status, with the number of processor bus break channels.

15.5   Break Functions

The TX49 debug support unit provides the following break functions:

- Instruction address break function.
- Data bus break function.
- Processor bus break function.

The instruction address break function monitors the program counter of the TX49 core and makes debug interrupt or trace trigger occur on the fixed virtual address via the following accesses:

- Specify the address to make a break occur in the IBA0 register.
- Use the IBM0 register to specify to each bit whether to compare the address in the IBA0 register.
- Use the IBC0 register to control the debug interrupt or trace trigger occurrence.
- Use the IBS register to check whether an instruction address break occurs.

Data bus break function monitors the interface between the execution unit of the TX49 core and level 1 cache and causes a debug interrupt or trace trigger to occur on the fixed virtual address or for the data via the following accesses:

- Specify the address or data to make a break occur in the DBA0 register or the DB0 register.
- Use the DBAM0 register to specify bits to compare against the address set in the DBA0 register.
- Use the DBC0 register to specify which bytes to compare to data.
- Use the DBC0 register to control the debug interrupt or trace trigger occurrence.
- Use the DBS register to check whether a data bus break occurs.

The processor bus break function monitors the interface to the TX49 core and makes a debug interrupt or trace trigger occur on the fixed virtual address or for data via the following accesses:

- Specify the address to make a break occur in the PBA0 register and the data to cause a breakpoint in the PBD0 register.
- Use the PBM0 register to specify whether to compare each bit set in the PBD0 register.
- Use the PBC0 register to specify to each bit whether to compare the address. Use the PBC0 register to control the debug interrupt or trace trigger occurrence.
- Use the PBC0 register to specify whether to make a break occur depending on the type of bus access (instruction fetch, data access, or cached/uncached area).
- Use the PBS register to check whether an instruction address break occurs.

15.6   Output by PC Trace

For real-time PC trace output in PC trace mode, the no-sequential program counter and trace information are output to GTPC [3:0] and GPCST [8:0] at 1/3 of the processor clock. The pipeline status of three clocks is output in GPCST by the processor clock. GPCST[8:6] is the first (the oldest) status, and GPCST[2:0] is the last (the latest) status.

Program counter values and exception codes are output in GTPC for every 4 bits. In order to decrease the number of pins of trace signals, the program counter values are output to GTPC only when program counter values are changed nonsequentially (indicating that a jump instruction is executed). The program counter to be output is of 30-bit or 44-bit length, which is selected by the PCLen bit in the JTAG_Control_Register. When the program counter to be output is 30-bit length, 8 cycles of DCLK (24 cycles of processor CLK) are required to output all 30 bits. When the program counter to be output is 44-bit length, 11 cycles of DCLK (33 cycles of processor CLK) are required to output all 44 bits. Because the exception code is 4-bit length, 1 cycle of DCLK (3 cycles of processor CLK) is required.

If the next jump instruction is generated before the program counter output by the past jump instruction is completed, one can choose to force termination of the past program counter output or to complete the past program counter output by stalling the pipeline. Select either one with the TM bit in the debug control register:

GPCST shows the pipeline status.

111 STL The pipeline is stalled.

110 JMP The jump instruction with the target PC output is executed (the target PC is output in GTPC).

101 BRT The jump instruction without the target PC output is executed.

100 EXP An exception occurs (the exception code is output in GTPC).

011 SEQ A normal instruction that is not a jump instruction is executed (including the case that the jump conditions are not met by the condition jump instruction).

010 TST A trace trigger occurs during pipeline stall.

001 TSQ A trace trigger occurs during execution of the normal instruction in the pipeline.

000 DBM The core is in debug mode (DM = 1).

Index